FIELD TECHNIQUES AND RESEARCH METHODS IN GEOGRAPHY

Robert H. Stoddard
Department of Geography
University of Nebraska
Lincoln, Nebraska

**KENDALL/HUNT
PUBLISHING COMPANY**
Dubuque, Iowa

**SERIES: National Council for Geographic Education
Pacesetter Series**

Clyde F. Kohn, Consulting Editor
The University of Iowa

Elizabeth Purdum, Manuscript Editor
The Florida State University

Cartography by The Florida
Resources and Environmental
Analysis Center, The Florida State University

Lounsbury/Sommers/Fernald: LAND USE: A Spatial Approach

Manson/Ridd: NEW PERSPECTIVES ON GEOGRAPHIC EDUCATION:
Putting Theory Into Practice

Richason: INTRODUCTION TO REMOTE SENSING OF THE
ENVIRONMENT

Richason: LABORATORY MANUAL FOR INTRODUCTION TO
REMOTE SENSING OF THE ENVIRONMENT

FIELD TECHNIQUES AND RESEARCH METHODS IN GEOGRAPHY
(Reviewers)

Professor Kenneth E. Corey—University of Maryland, College Park

Professor John W. Frazier—State University of New York, Binghamton

Professor Rutherford H. Platt—University of Massachusetts, Boston

Professor Otto F. Jakubek—Central Washington University

Copyright © 1982 by National Council for Geographic Education

Library of Congress Catalog Card Number: 81–84843

ISBN 0–8403–2602–5

Printed in the United States of America

C   402602   01

Contents

Preface

Inexperienced researchers seeking to solve geographic problems are often confronted with a wide choice of techniques available for acquiring data. Unfortunately, information about these techniques is scattered among various publications. This text was written, therefore, for the purpose of assembling several potential field methods into one publication. It provides aspiring researchers with the opportunity to compare alternative types of data-gathering techniques and to select those appropriate to particular research problems.

Although the emphasis is on research in human geography, the techniques presented are useful for other groups of readers such as urban planners, regional planners, sociologists, and others concerned with areal data and locational problems. Some terminology, several cited references, and the suggested applications at the end of chapters are intended for a typical college junior, senior, or beginning graduate student in geography. Nevertheless, the subject matter should be meaningful to and usable by persons in other disciplines and those not formally enrolled in school, such as community volunteers, agency personnel, and teachers of the social sciences. For readers unfamiliar with the geographic perspective, Chapter 1 and the numerous applications to the hypothetical city of Choros should provide an adequate background for understanding the remainder of the text.

In places the text may seem elementary for readers who already know something about the nature of sampling, ways of designing a questionnaire, or strategies for preventing bias. I agree that many techniques are fairly simple to perform and explanations about procedures may appear rudimentary, but the apparent simplicity of procedures may lull the amateur researcher into overlooking the myriad ways data can become distorted. A major theme in the books warns against using a deceptively easy technique without evaluating the many factors that may produce inaccurate results. For example, a questionnaire that appeared in a local newspaper a few years ago illustrates the kind of difficulty that may occur in the commonly used technique of "just asking a few questions." Respondents were asked to check one of the four following choices to this statement:

Today's children would be better behaved if—

    a. Home discipline were stronger.
    b. Home discipline were more lenient.
    c. Fathers took more responsibility for discipline.
    d. Mothers stayed home more frequently.

Given this set of choices, how could the respondents indicate their beliefs accurately? After obtaining the replies, how could the author make a valid interpretation of the intended responses? The goal of this book is to assist readers in detecting and preventing the kind of errors typified by that newspaper questionnaire.

I do not claim that every technique used in geographic research is included in this text, but it is hoped that this survey of major methods will serve as a guide so that the aspiring researcher can evaluate field techniques critically and consequently devise original strategies appropriate for a particular research problem. Also, even though the illustrations and references pertain to data in the United States, most techniques are applicable to any location and are not limited to a single region.

Background information about the role of field data is presented in Chapter 1. Chapters 2, 3, 4, and 10 deal with the many preparatory tasks and decisions that precede the actual collection of data. The remaining chapters describe and evaluate various groups of data-collecting techniques. One group of techniques are those requiring direct visual observations of human beings, and of their cultural and natural environments (Chapter 5). Another group deals with strategies that obtain data by asking persons questions about themselves and their spatial activities (Chapter 6). A third group concerns the data that are available from the imagery produced by remote sensors (Chapter 7). A fourth group concerns information acquired by consulting primary sources which may be found in libraries, data banks, attics, and other places where archival items and similar materials may be stored (Chapter 8). One chapter (Chapter 9) pertains primarily to the difficulties encountered in observing and measuring phenomena in motion.

Techniques that are similar tend to be grouped together in chapters so the reader can find comments about broad classes of techniques in one section. However, the many variations in, and overlapping of, techniques make the chapter divisions rather arbitrary. Furthermore, discussions which pertain to all techniques such as bias and the evaluative check-list of Table 5.2 are not repeated in each chapter. For maximum benefit, therefore, the reader should read the chapters sequentially and/or follow the cross-references.

Persons who voluntarily read an early draft of the text and offered helpful reactions are Thomas D. Anderson, Donald W. Buckwalter, William A. Dando, Richard E. Lonsdale, Harold J. Retallick, and Donald C. Rundquist. In addition, Justin C. Friberg, Percy (Doc) H. Dougherty, and Timothy J. Rickard donated considerable time by providing many valuable suggestions. At a later stage, Clyde F. Kohn contributed various editorial improvements. I thank them all for their assistance but certainly do not hold them accountable for the limitations that remain in the text.

I am especially grateful to my wife, Sally Stoddard. Her role was not just the usual "she-kept-the-children-quiet-while-father-wrote" type of activity. As an experienced linguist, she toiled through multiple versions of the manuscript and made many stylistic suggestions. Her perspective as a non-geographer often helped convert an ambiguous statement into a much clearer message.

Robert H. Stoddard

Chapter 1

Introduction: Collecting Areal Data

News in the small city of Choros is usually about the weather, the success of the high school basketball team, or events sponsored by churches and other organizations. Recently, though, the news in the local paper and the talk around town—in the taverns, on the streets, and at Marie's Cafe—has focused on the controversial "school building issue." This controversy, which has become quite heated, is basically about the location of a new elementary school.

The school building issue arose because the existing elementary school building has become unsafe with age and inadequate with the increasing population of the city. The present structure must be replaced by greatly renovating the existing building or constructing one or more new buildings. The major dispute concerns the location of the building(s). Proposed locations are the following: (1) the site of the existing building, either for its renovation or for a new structure, (2) an unspecified site somewhere in the southeast portion of the city (see Area F in fig. 1.1), and (3) two sites in different sections of the city where two medium sized buildings could be constructed (Areas G_1 and G_2 in fig. 1.1).

Although the fundamental problem deals with the question of location, the controversy erupted over a wide range of interrelated—and sometimes not so related—questions. Typically citizens ask: "How much will it raise my taxes?" "Why should my child have to go so far to a new school?" "Why can't those rich folks with two cars drive their kids to school?" "How are the 'little ones' going to get across Highway #10?" "Who's the guy who'll make a killing by selling his land for a new school?" "Can't they see that the present school is in a terrible location so the new building ought to be elsewhere?" "How come people in 'that' part of town are such troublemakers?"

Even though some people act as if the volume of their shouted arguments or the intensity of their polemics will produce converts, most persons realize that facts about the existing conditions are necessary for making convincing arguments and valid conclusions. Some people frequently behave on the basis of emotions, traditions, and faith; others more often base their decisions on empirical facts about their environment. The dependence of the latter group on environmental facts in turn means they must be able to rely on the methods by which those facts are acquired. Therefore, if the citizens of Choros want to solve their problem on the basis of meaningful data, they need to consider both the kind of data required and the suitability of techniques used to gather those data.

Even though Choros citizens may believe their problem is especially critical, this type of locational controversy is not unique. In general, as the human population increases and becomes more mobile, the number of interpersonal contacts with their accompanying conflicts increases. To find satisfactory solutions to societal conflicts, problem-solvers need data—data that are accurate and pertinent to the issues in contention. Consequently, the demand for data that are obtained by reliable methods is expected to expand along with the increase in human interactions.

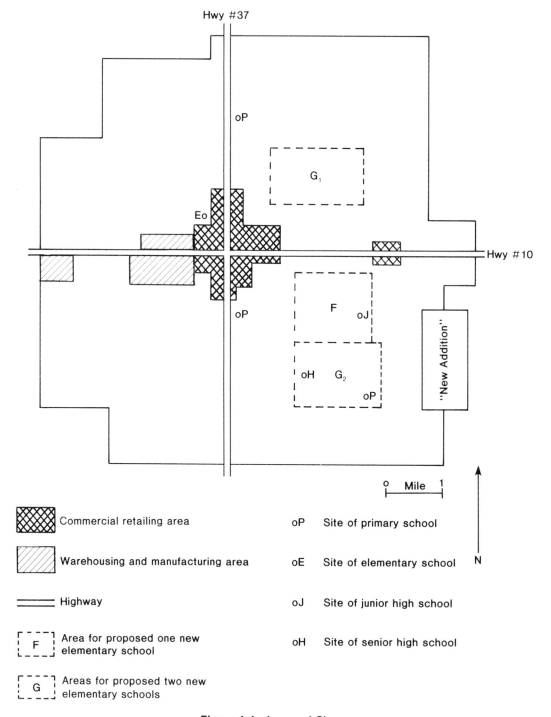

Figure 1.1. A map of Choros

The purpose of this text is to deal with the need for collecting useful data to solve societal problems. The school building issue, which concerns the citizens of Choros, typifies many problems confronting society. Its specific solution may require only a few field techniques selected from among many, yet it can serve as an illustration for several aspects of data collection. Therefore, Choros is mentioned frequently in this text as an example of the kind of questions confronting anyone who seeks to gather original data.

This first chapter concentrates primarily on the following questions:

What kinds of problems are especially geographic in their perspective?
What are the methodological reasons for gathering geographic data?
What are some general characteristics of field data?
How does data collection fit into the total methodology of solving a research problem?
Why are various field techniques presented in the order they appear in this book?

A Geographic Perspective

Locating the school building is essentially a *geographic problem,* even though some persons may not think of it as primarily geographic. Whether a person regards a problem as "geographic" or not is not too important as long as people are able to communicate about the subject. (A rose by any other name may smell as sweet—but a person would have difficulty ordering a rose over the telephone if florists did not use the same name as purchasers). Even though some readers may not regard themselves as geographers, they should understand the meaning and distinctive characteristic of "a geographic problem."

The term "problem" is applied to an intellectual stimulus calling for a response in the form of a solution. In this text solutions are based on relationships which can be revealed and scrutinized by using the scientific methodology. This method provides an effective basis for solving problems when groups of different things relate to and interact with each other.

The phenomena which affect human lives and generate uncertainties are numerous and often complex, so scholars usually divide the academic responsibilities according to the kind of questions asked. Some scholars (chemists) deal with the structure and processes of the chemical world; others (sociologists) inquire about the characteristics and behavior of social groups; and still others (geologists) study the conditions that provide variations in the materials of the earth. These disciplinary distinctions are not made because the real world consists of many isolated parts, but rather, because humans find it necessary to deal with a limited set of questions at a time. Through an academic division of labor each group of scholars can concentrate on a specific set of questions and methods for solving them.

Characteristics of Geography

The primary questions that *geographers* attempt to answer cluster around those involving *where* and *why there.* That is, geographers search for understanding about the locations of phenomena; they seek an explanation for the areal arrangements and spatial relationships of phenomena on the earth's surface. By concentrating upon the aspects of location, geographers strive to provide knowledge about one critical facet of humanity's total environment.

This specialization by geographers on the *spatial aspects* of phenomena does not mean that other scientists exclude location from their investigations. Every event occurs at a specific place and time, so these facts might become a part of the information observed by any scholar. Nevertheless, many scientists attempt to formulate statements about relationships that apply irrespective of their spatial and temporal occurrences. Normally the chemist purposely attempts to study the relationships of chemical phenomena in controlled laboratory conditions exclusive of their spatial-temporal setting. Other scholars may include location but only as one of several factors influencing the behavior of a specified phenomenon. For example, a sociologist who attempts to understand the importance of the family in society may note that location affects the role of family, but that factor is probably considered as only one of several attributes. Likewise, a geologist may or may not include the spatial positions of volcanoes when analyzing the origins and processes of vulcanism.

The geographer is *always* and *primarily* interested in location. Relationships that apply when location is excluded, such as a principle of chemical behavior, are secondary for the geographer. Those relationships that *do* include location are of primary concern to geographers, but from a different perspective than emphasized by non-geographers. Whereas the non-geographer may seek knowledge by including location as one of several phenomena related to a particular problem, the geographer seeks understanding about the locations of phenomena because such knowledge contributes directly to learning about humans and their environments.

To solve geographic problems, therefore, a person must obtain information about the locations of those things which are thought to be related to possible solutions. Usually spatial analysis is based on information about a set of places, with the data pertaining to each individual position constituting a *geographic fact*. The meaning of this expression refers to a fact, which is an empirically verifiable statement. The fact is geographic if it associates a specific phenomenon with a particular earth location. For example, the first four statements that follow illustrate geographic facts, but the fifth one is not geographic in its present form:

> In 1946 the largest cattle ranch was in Carmen County.
> A rock band performed on the mall every Friday during the summer.
> Mecca is the site of a famous pilgrimage center.
> This intersection is the site of the latest traffic fatality.
> The temperature on the 4th of July was more than 27°C.

The four geographic facts not only illustrate the importance of location but they also exemplify another characteristic of geography, namely, its involvement with a wide variety of phenomena. Many disciplines tend to concentrate on a limited range of phenomena (e.g., the chemical aspects of matter; the sociological groupings of humans; the geological features of the earth), but geography may deal with *any* phenomenon (e.g., pilgrimage centers, rock bands, glaciated limestone beds, persons suffering from agoraphobia, artesian wells, and the proverbial kitchen sink). A major effect of this large diversity of topics is the disparity in techniques for gathering data, a characteristic that is manifested in the variety of techniques described in this book.

In summary, a geographic problem involves an inquiry about reasons for the location of some phenomenon on the earth. The phenomenon may be anything: a single unit or a set of elements, tangible or intangible. Solutions to the problems are sought by finding spatial relationships among explanatory phenomena. Successful answers normally explain why one phenomenon is located where it is in terms of its position relative to the positions of other phenomena.[1]

The school building issue in Choros is essentially a geographic problem because it deals with location. The ultimate goal is to decide the best place to locate the elementary school, whether by renovation at the present site or by building at one or even two new locations. To reach that decision, information must be obtained about the spatial patterns of school children, land values, traffic, citizens' attitudes toward education, and other phenomena related to the school siting. The reason the Choros problem is termed "geographic" is not because it pertains to a particular kind of phenomenon (e.g., the erroneous impression by many non-geographers that "geographers always study the weather") but rather because it involves studying locations and spatial relationships.

Somewhat contrasting views of the Choros issue and the role of geography are expressed by three individuals who are concerned about the location of the school building. Ms. S. is a university student who is doing a research project on the location of educational services in small cities. She decided to conduct field research in Choros because it is only fifty miles away from the university and because the city's plans to build a new school permit her to examine spatial relations that are very current. As a geography major she is already aware of the importance of location in the problem, so she wants to learn as much as possible about where the school site will be located relative to other features of the city, and how the citizens of Choros make that decision.

Dr. A. is the top administrator of the elementary school program in Choros. She has observed the deteriorating condition of the old school building for several years and has been urging citizens to build new facilities. She is glad that the school board is committed to a new building now and that definite proposals are being considered seriously by the community. Dr. A. did not specialize in geography for her academic degree, and she does not consider the school building problem as one to be analyzed only by geographers. She does recognize, however, that many parents are concerned about the location of the new school relative to their homes and, particularly, to the traffic on Highway #10.

Mr. C. is a citizen living near the existing school site. He has become quite disturbed about the possible cost of a new building (or, worse yet, possibly two new buildings) located on "new" school land. He has formed an ad hoc group to advocate remodeling the existing building and to resist all proposed sites located anywhere other than on the existing school land. He is sure that by collecting "all the facts" he can prove to the taxpayers of Choros that there is only one sensible location for the elementary school. Mr. C. disclaims any expertise as a geographer. In fact, he does not profess any specialty ("I'm just an ordinary, law-abiding citizen"). However his desire to show others that a certain location is the best one, is certainly very geographic.

Although these three individuals may view the discipline called geography differently, they all are involved with a problem that is locational. How they approach the problem and collect data is pertinent for this text. Likewise, irrespective of whether you, the reader, regard yourself as a geographer or not, the techniques presented here are applicable whenever you need to gather data to solve problems that take location into account.

In this text, a *researcher* refers to anyone who gathers data and utilizes those facts to gain knowledge and to solve problems. A researcher may be a scholar studying formally in a school like Ms. S., a public employee working in a governmental organization like Dr. A., or a citizen volunteering to help with a community like Mr. C. Because new problems often involve somewhat different conditions from those of the past, even experienced scholars must continually learn about the best ways of collecting data. The terms "amateur" and "novice," therefore, should not be interpreted as derogatory but only as an indication of a researcher who has had limited training and experience with field techniques.

Various Approaches to Geographic Problems

The approaches to the Choros problem by these three persons illustrates methodological contrasts that are critical to various field techniques. To comprehend some of the issues that are discussed in subsequent chapters, the reader needs to be aware of several approaches to geographic problems. Ms. S., who is the student conducting research on the location of educational service in small cities, is interested in *developing general principles of location.*[2] She considers Choros a typical small city and she assumes that her results represent conditions in many other cities with similar characteristics. She will collect several sets of data and analyze them for their overall trends and relationships. By this method she will generalize her conclusions from the observed patterns in the specific area of Choros to a larger group of similar cities. Because this approach is the normal one for most scientific research, the establishment of locational principles is given the greatest emphasis in this text.

Dr. A. approaches the Choros problem differently because she accepts the principles of location established previously and plans to apply them to the specific Choros problem. Her approach, which typifies *applied geography,* contrasts with the one followed by Ms. S. because Dr. A. will commence with a generalization and then apply it to a specific case.

A third approach is typified by Mr. C. who commenced with the "conclusion" that the best location for the school building is the site of the existing building. He plans to ascertain information about citizen support for minimizing building costs. He will search for data about financial advantages of using land already owned by the school district and of renovating the existing building. The primary fault with the approach by Mr. C. is that he begins with a specific locational answer and plans to find data that will support his conclusion. This kind of reasoning is done frequently when people rationalize their prior decisions, but it will not necessarily produce the best answer from a choice of several options. Therefore, the overall research method of Mr. C. is not pursued in subsequent sections, even though some of his data-gathering techniques are reported for illustrative purposes.

Two additional approaches to geographic knowledge are those associated with *theoretical geography* and *descriptive geography.* A theoretician who attempts to solve a spatial problem by creating a general theory of locations builds on assumptions, definitions, and postulates. The logic of the resulting theoretical system does not depend on empirical data. Although the conclusions deduced by a theoretical study are very helpful in establishing general principles of location and the utility of a locational theory is usually judged by its verification in the "real" world, a geographic theory does not require the kind of field data discussed in this book.

The descriptive approach refers to the acquisition of many facts about "what is where," primarily for the purpose of accumulating information about the particular phenomena located in a specific area. Articles in the *National Geographic* magazine, travel books for tourists, gazetteers, and world almanacs are examples of descriptive geographies which have the goal of telling readers about what phenomena occur in identified places in the world. The major purpose for obtaining these data is to collect a multitude of geographic facts, which can be conveyed to curious readers. The intent is not to solve a problem that requires geographic input for its solution. Although the usefulness of most descriptive accounts, as for example, those in a travel guide, is dependent on accurate and current data, the descriptive approach is not emphasized in this text because it can be subsumed under the problem-solving approach. To solve a locational problem requires the collection of data similar to that which is obtained for a descriptive account, but the obligation to analyze a defined geographic problem gives greater direction and focus to the data-gathering procedures.

In summary, geographic studies may be approached by any one of four different methodologies. A geographer doing a descriptive study (Type I in Table 1.1) observes "what is where" in the empirical world and then reports those facts in the form of maps, pictures, and prose. A geographer attempting to develop principles of location (Type II in Table 1.1) observes the empirical world to learn what phenomena are located where, especially in terms of their relative positions, and then generalizes these locational relationships (e.g., where Xs are found, Ys tend to be nearby). A geographer wanting to apply spatial knowledge (Type III) studies the geographic principles established by previous scholars and also observes the existing locations of phenomena in a particular area. The principles are then applied to the observed locational facts to predict future patterns or to prescribe how selected phenomena should be arranged to achieve specified objectives. A geographer engaged in a theoretical study (type IV) uses deductive reasoning to predict where certain phenomena would be located under specified conditions. The emphasis in this book is placed primarily on developing geographic principles (Type II).

Table 1.1   Approaches to Geographic Studies

Type of Approach	Activity	Results
I Describing Places	Observing what is where in the empirical world by absolute locations	Descriptions and Representations of What is Where
II Developing Geographic Principles	Observing what is where in the empirical world by relative positions	Generalizations of What is Where
III Applying Geographic Principles	Observing what is where in the empirical world and Applying Geographic Principles	Predictions or Prescriptions of What will/should be Where
IV Theorizing Geographic Relationships	Deducing what would be where based on logical reasoning	Predictions of What would be Where under Specified Conditions

Nature of Data

The data sought are geographic facts. These possess some differentiating characteristics, mostly because of their association within a complex phenomena located on the areal surface of the earth. The nature of geographic data can be examined by noting that they (1) pertain to phenomena which occur over an area, (2) are collected in the field, and (3) involve many kinds of phenomena.

Areal Data

Geographic facts are distinguished from non-spatial data in at least two ways. These distinctions (which may cause difficulties for a researcher attempting to collect locational information) arise from their dimensionality and the lack of individuality of some phenomena.

Dimensions

Geography deals with the locations of phenomena on the surface of the earth, which means that every geographic fact must possess a minimum of two data elements, namely, the coordinates of two-dimensional space. To designate a location with a single name is not an exception because the name in itself, without an accompanying map, lacks locational meaning (see more below.) These spatial elements represent two additional data elements that are required for geographic facts but which are not necessary for non-spatial studies. A geographer who wants information about m attributes per object, therefore, must collect $m + 2$ facts for each object, where m symbolizes a variable number.

This is illustrated by some measurements taken by Ms. S. of noise levels in Choros. She organized her measurements in six columns with headings as follows: (1) location, distance from Highway #10; (2) location, distance from Highway #37; (3) time of day; (4) wind direction; (5) wind velocity; and (6) noise level. Thus, in columns 3 through 6 she recorded four data facts ($m = 4$) about phenomena probably related to noise; but because she wanted to analyze how they varied geographically, she also had to include the two locational facts (columns 1 and 2) for each observation site.

Although most geographic problems are considered in two-dimensional space on the surface of the earth, a few special kinds of phenomena should be measured according to their locations in three-dimensional space. Ore beds, atmospheric conditions, and uses of floor space in multi-storied buildings are phenomena that might be located by earth position plus depth or height. This means that these geographic facts require three, not just two, data elements for each individual as well as the other attributes being studied.

Individual Geographic Units

A second distinctive characteristic of areal data is the lack of individuality for some phenomena. The word *individual* refers to a single object, person, or event, such as a highway intersection, a shoe factory, a Choros student, a college campus, a performance by a rock band, or a kitchen sink. The reason for this more inclusive meaning, which contrasts to a definition restricted to only human beings, involves the topic of sampling (Chapter 4); but here it is important to realize that the term "individual" may even pertain to a single unit of earth space.

Individuals that occur at distinct locations, separated by nonoccurrences, i.e., areas where the phenomenon does not exist, constitute what is called a geographically *discrete* phenomenon. Although each individual of a discrete phenomenon in reality occupies a small area of the earth's surface, each location is regarded as a point (or, a geometrically non-dimensional occurrence). Dot symbols on a state map could represent the locations of discrete phenomena such as shoe factories and college campuses. Thus, the task of specifying the locations of individuals which are geographically discrete consists only of measuring the position of each point.

In contrast to discrete phenomena, those phenomena that occur throughout a study area are termed geographically *continuous* phenomena. Thus, continuous phenomena, because they exist everywhere, do *not* consist of individual geographic units per se. They possess geographic interest because they vary from place to place. Places differ; but where does one "place" stop and another begin? What is the areal extent of each individual unit of area? The answer is provided by each researcher who decides that an area of arbitrary size will constitute an individual, which may be called the *minimum areal unit*. Whether the researcher designates a square meter, a city block, a hectare, or some other areal unit as the minimum size depends on a variety of research considerations (discussed in later chapters).

Two additional comments should be made about areally discrete and continuous phenomena. First, the distinction between the two kinds of distribution is partly a function of scale. A spatially detailed study of a college campus might examine the variations in the frequency of lawn use from place to place, and thus, be a study of a continuous phenomenon. In this case the area is small and would be shown cartographically on a large-scale map. If the scale of study were shifted to using a small-scale map so areas previously examined for internal variations shrink to homogeneous points, then the college campuses would be studied as discrete occurrences.

Secondly, the distinction between geographically discrete and continuous phenomena refers to their characteristics in terms of geometry. The terms discrete and continuous are used also to refer to the kind of measurements made to produce non-spatial data. Measurements that produce numerical data which are discrete are those that exclude certain values, often fractions or non-integers. For example, the number of children attending Choros schools must be an integer, such as 4237—it cannot be a quantity measured as 4237.8712. In contrast, data produced by measurements that include continuous values are not restricted to only integers. To illustrate, the quantity of energy consumed per month in the Choros schools may be measured in units for which a number such as 4237.8712 is quite meaningful.

Field Data

Geographic facts pertain to phenomena occurring within their earthbound setting, so the data must be extracted from a complex of interrelated phenomena. The features of interest cannot be carried into a laboratory for analysis but rather must be observed within the context of their environment. Indeed, the term *field* usually refers to the contextual setting of phenomena, in contrast to a laboratory setting where the observed phenomena can be isolated from many external factors. Collecting field data is complicated, therefore, by the necessity of extracting the data about the relevant phenomena from their setting within an interactive environmental system.

Many field data are obtained outdoors, for example, when land uses are observed or where people are going about their normal activities and are asked questions by researchers. In addition, some scholars may apply the term field data to information extracted from stored documents such as aerial photographs and from the diaries of pioneer travelers. The use of the term "field" should be applied to stored materials, however, only when new information is obtained through a purposeful extraction of selected items (Chapter 8). It should not pertain to using secondary library sources. In this text, field data refer to this broader interpretation. A comprehensive coverage of many techniques and strategies used to obtain original geographic data is presented.[3]

Kinds of Phenomena

Even though any earth-bound phenomenon can be studied spatially, individual geographers often specialize by restricting the number of phenomena they study. They may concentrate on any combination of phenomena, but there is a tendency to group phenomena into traditional classes. At the broadest level of classes, geographers separate phenomena that occur in nature (physical geography) from those made by, and consisting of, humans (human/cultural geography). This division may confuse persons not familiar with geographic traditions because some features, such as buildings, could be called physical since they are tangible objects, whereas geographers would normally class them as cultural (i.e., non-physical) because they are manmade.

This possible confusion about the meaning of "physical" features suggests that a different classification of phenomena might be helpful. Another reason for breaking with tradition, however, directly affects the classification of field techniques. The appropriateness of a particular data gathering technique over an alternative one is related to the degree that humans can control the phenomenon being observed. Specifically, techniques used to gather data about an inanimate object often differ greatly from those utilized to obtain information from a person. Thus, a three-class division of phenomena is adopted for this text: natural features (N), cultural features (C), and people (P). The first class consists of those phenomena that are primarily a part of nature; the second are made and controlled by human beings; and the third are the people themselves.

The visibility of the phenomenon is another attribute of phenomena that affects the suitability of various field techniques. Phenomena may be classified, therefore, according to whether they are visible (V) or invisible (I). When these two criteria are combined, six classes result (fig. 1.2).

This six-class scheme does not eliminate all classificatory and communication difficulties because there are still some transitional elements. For instance, some soils exist primarily in their natural state but others have been greatly altered by humans; so "soils" might belong partly in class NV and partly in CV. By definition crops are controlled by humans and therefore could be classed as CV, yet they also constitute part of the botanical world of nature (i.e., class NV). Similarly, atmospheric temperature may be invisible (i.e., class NI), but certain forms of precipitation are so highly correlated with temperature that people think they "see" the coldness of a snow-filled wind. Furthermore, certain conditions (e.g., temperature of air surrounding a house) may not be seen by the human eye so might be classed NI, but they can be sensed by an infrared camera and subsequently seen in a picture, which would push them into a visible category.

Data about nature phenomena (N) are frequently collected by observing them visually or by measuring them with instruments. Techniques for making visible observations of phenonena (NI, CI, and PI) are discussed in Chapters 3 and 5. Instruments designed for measuring other

	Natural Phenomena (N)	"Cultural" Phenomena (C)	People (P)
Visible (V)	Landforms	Buildings	Pedestrians
	Soils		
	Crops		
Invisible (I)	Atmospheric temperatures	Urban noises	Individual attitudes

Figure 1.2. Classes of phenomena, with illustrative topics

characteristics of natural phenomena are usually quite specialized, so techniques for their utilization require rather specific training. It would take a voluminous manual to describe the operation of all the different instruments that might be used to measure all natural phenomena. Consequently, most techniques that involve the use of instruments to measure natural phenomena (N) are omitted from this text. The major exception is a section on utilizing data obtained from remote sensors (Chapter 7). Stated otherwise, the text concentrates on field techniques that are more applicable for research about cultural phenomena and people than about natural features.

Research Methodology

The appropriateness of various techniques is partly related to the nature of the data and of the phenomena being observed. It is also related to the total research methodology. For methodologies designed to establish locational principles (the Type II approach) or to solve applied problems (Type III), the research objective greatly influences the kind of data needed and, hence, the choice of an appropriate technique for obtaining the data. This section, therefore, briefly considers the standard scientific research methodology, associated particularly with the Type II approach, and the place of data collection in the total research process.

People find answers to their questions about life and their environment in a variety of ways. Many questions involve feelings and attitudes, and these are answered in terms of emotions. Other questions about life, death, and the meaning of existence may be answered primarily with affirmations of faith. Still other questions are solved best by following a scientific procedure. Various kinds of questions elicit answers from different domains, such as the emotional, theological, and scientific domains. The appropriate domain depends on the nature of the query.

Most questions raised by the problems of geography fall within the scientific domain; so the proper methodology for solving these problems is a scientific one. The traditional and dominant

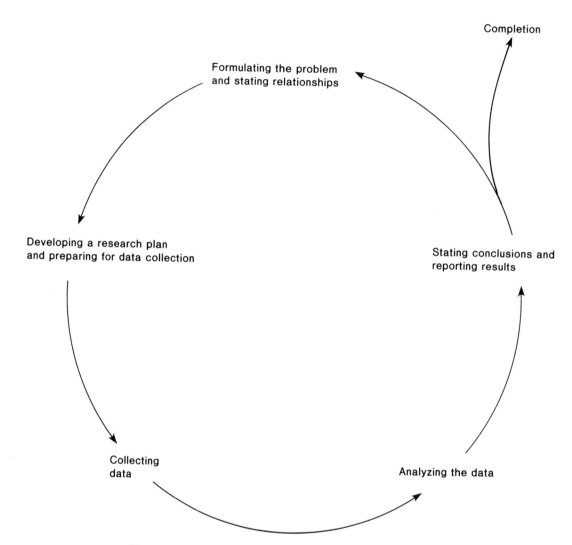

Figure 1.3. Stages in the cycle of a standard research plan

model for scientific research is the one that stresses collecting data by objective measurements and making conclusions on the basis of relationships that have been established mathematically. Some geographers and other social scientists utilize alternative models that attempt to gain understanding through more consciously subjective models (e.g., the phenomenological one).[4] Although experiential fieldwork and similar techniques are discussed in this text (see Chapter 5), the emphasis is on gathering data for what is commonly called "objective" analysis.

The nature of the scientific methodology is probably familiar to most readers; so it is unnecessary to describe the procedure in detail.[5] However, the major stages of the research cycle should be reviewed for their relevance to collecting field data (fig. 1.3).

Formulating the Problem and Stating Relationships

Entry into the cycle of research stages normally commences with a problem. Most problems provoke curiosity, expose a gap in the body of knowledge about some phenomenon, or constitute a societal ill. In the case of the school building issue the problem is expressed by the citizens by the question: "Where should we locate our new elementary school building?" Next, some possible answers are suggested on the basis of known or expected relationships. In Choros the proposed solutions are at the locations identified as E, F, or G_1 and G_2 (fig. 1.1). These two aspects of research can be formalized by a statement of the problem and a set of hypotheses or generalizations.

Statement of Problem

Usually the primary research problem is worded as a simple statement that expresses the core of the research. The statement invariably mentions the phenomenon being studied, which may be termed the *research phenomenon* or *problem variable*. The problem variable, therefore, is one of the phenomena that must be measured by an appropriate field technique. For example, Ms. S. stated her research problem as this: "The purpose of the study is to determine the spatial relationships between school buildings and selected socio-economic phenomena in a small city in the late 1970s." The research phenomenon in her case is the set of all school buildings in Choros: three primary schools, the junior high school, the senior high school, the existing elementary building, and the site(s) of any new building(s). Her statement does not specify the "socio-economic phenomena," so a formal set of hypotheses or a subsidiary statement that gives more detail will need to accompany this somewhat vague statement.

Dr. A., who is approaching the problem from an applied viewpoint, expresses the geographic portion of the problem like this: "The problem is to find the best location for the new elementary school building(s) in terms of the location and size of city neighborhoods, the accessibility of pupils, and traffic patterns." According to her statement, the problem variable consists of all geographic positions in Choros, but in reality, the possible choices have been narrowed to site E, area F, and the G areas. The other phenomena, which are those she accepts as being pertinent for solving the locational problem, require data about neighborhoods, accessibility, and traffic patterns.

Hypotheses and Other Relationships

Hypotheses are educated guesses (or, proposed generalizations) about the relationships that are thought to exist between the research phenomenon and the other pertinent phenomena which are frequently termed *explanatory variables* (also, predictor or independent variables). These variables are the ones about which data must be gathered; so they serve as a direct guide to the prospective field worker. Furthermore, the relationship that is stated in each hypothesis reveals the particular attributes of each explanatory phenomenon that must be measured during the data gathering stage. If, for example, Ms. S. were to hypothesize "that the future school building will be placed at the point of minimum aggregate travel for the elementary student population," the precision of her statement would serve as an excellent guide for knowing what data to collect.

Sometimes scholars, even when following the general procedures for establishing geographic principles (by the Type II approach), do not present hypotheses at the beginning of their research. By postponing a decision about potential relationships, researchers essentially shift some major questions about data needs. If no hypotheses are stated or relationships are only vaguely implied, important decisions about explanatory variables are, in effect, delayed until the stage of data collection.

When research is undertaken for an applied problem (the Type III approach), the terminology varies somewhat. Because the researcher commences with a generalization that is already accepted, the "hypothesis" is replaced by a "principle." For example, Dr. A. also made a statement of relationship, which took the form: The best location for an elementary school is near the center of a residential area identified as a neighborhood. As you can see, this statement is somewhat less precise about spatial relationships than the one by Ms. S.

The degree of specificity, whether stated as a working hypothesis or an accepted principle, affects the timing of certain research decisions. The researcher must make the decision, whether early or late, about which variables are relevant and how they are assumed to interrelate. Aspiring researchers who procrastinate too long may find themselves standing in the field wondering what data to collect.

Developing a Research Plan and Preparing for Data Collection

The decision about which phenomena are meaningful is only one of several decisions that must be made prior to collecting data. Because the several decisions are interdependent, it is wise to organize these decisions into a *research plan*. This is a working paper that guides the researcher as decisions about definitions and procedures are made. It is truly an ongoing working guide because it is started at the beginning of a research project and continued until after a field reconnaissance has been completed (Chapter 2). Other decisions are discussed in the next three chapters, and the elements of a complete research plan are described in Chapter 10.

One kind of decision that needs to be made during early stages of research is the translation of concepts into operational equivalents; that is, a problem is normally stated in terms of concepts that can be communicated intellectually to other persons, but these conceptual terms need to be converted into operational definitions. An *operational definition* bridges the gap between the conceptual level of the mind and the empirical level of observations by human senses.

More comments about operational definitions are presented in Chapter 2, but the kinds of questions that may arise are illustrated by the Choros controversy. Persons may talk about locating the new building "close to all the elementary kids." This general goal is understandable and its relative merits can be debated as a concept. A person is confronted by several questions, however, when attempting to apply this goal to specific conditions. How do you define and measure "close-ness"? Is this by earth distance (in miles or kilometers), or by time distance (in minutes of traveling time)? Is it by direct geometric distance between home and school, or is it by a traveled route along the streets? Should families with more than one elementary-age child who drive all their children to school in a single car be counted as engaging in more than one pupil-trip? Who is included as an "elementary pupil"? Should the children who are in the upper primary now but who will attend elementary classes next fall be part of the data set?

In addition to defining the variables being studied, a geographer must designate the area where these phenomena are to be observed. In the case of the school building the definition and delineation of the *study area* appears to be simplified by the existence of boundaries for the Choros school district. However, as explained in Chapters 2 and 4, acceptance of these school district boundaries carry some methodological implications that need to be considered carefully. Furthermore, a researcher does not always find established boundaries that can be used to delineate a study area, so questions about size and location of the area require answers.

Collecting the Data

The primary topic of this entire book pertains to techniques for the collection of data; thus, details about the stage of data collection are unnecessary here. It should be emphasized, however, that data are meaningful only as a mode for answering a specific research problem and are useful to the degree that they can be analyzed to provide a solution. Even though attention in other chapters is on techniques for gathering data, the reader must remember that the data produced by various techniques are valuable only as they satisfy overall research goals.

Analyzing the Data

The stage of analysis refers to the process by which a researcher organizes and summarizes a mass of figures and other data into a form that exposes an answer. This analytical process usually involves calculating summarizing statistics. Many forms of summaries (e.g., means, medians, frequency counts) are well known; thus, even amateur researchers can calculate and present them. Further analysis by more complex statistical methods may necessitate special training. Although data analysis is not the subject of this text, it should be noted that the use of progressively more precise analytical tools creates a demand for more accurate and precise data.

Another interrelationship between data collection and data analysis occurs because certain statistical methods are limited by the data characteristics. Generally, if the data consist of *numbers* obtained by measuring phenomena, then most statistics can be applied. In contrast, if the data result from classifying phenomena without numerical associations, then fewer kinds of statistical summaries are available (see more in Chapter 3).

Stating Conclusions and Reporting Results

In standard scientific methodology the formal testing of a hypothesis leads directly to a conclusion about its acceptability. Even a less formalized analytical procedure should produce evidence about the correctness of the hypotheses or the initial "guesses" about relationships between the research phenomenon and explanatory variables. The statement of acceptance or non-acceptance of each hypothesis, or less formalized guess, is important because it is the goal of the entire undertaking and because it constitutes the creative contribution of the researcher. In spite of the attention always given to the conclusion, all who acknowledge and accept it must also recognize that its validity is highly contingent on the quality of the supporting data. This again emphasizes the fact that the scientific method depends on the successful execution of all stages of the research cycle.

When the methodology for applied geography is followed, there is no rejection of a hypothesis or proposed generalization. The initial generalization is already accepted as a principle that is applicable to the research problem and study area. After the analytical stage is completed, the solution is known. For example, if Dr. A. accepts the generalization that the best location for a school in American cities is near the center of a neighborhood, then the answer is known once the data are collected and analyzed to determine the center of a neighborhood. Although the deductive reasoning of the applied approach contrasts with the inductive logic of the standard scientific methodology, both are highly dependent on valid empirical data for accurate conclusions.

The form in which conclusions are reported varies with the researcher, the intended audience, time constraints, and other conditions. Dr. A. might present her results as a written recommendation to the Choros School Board. Mr. C. probably will announce his conclusions to all who will listen in a variety of settings. Ms. S. may be required to write a scholarly paper that can be examined, challenged, and accepted by persons both within and outside the academic community. Although the forms may vary, a critical component of every report is the summary of the techniques used to obtain the data.

In some situations reaching a conclusion solves the problem and terminates the research project. For persons in Choros the final decision about the site (or, sites) for the new (or renovated) school building (or, buildings) ends their problem. In a different context, Ms. S. may conclude that one or more of her hypotheses are acceptable and can be generalized as a principle of location. Those hypotheses that are not accepted may expose new questions about the location of school buildings. Although she may not pursue these new uncertainties herself, she should report them in a form that other scholars interested in this subfield of knowledge can benefit from her study and can seek additional answers in a new research project. It is in this manner that the search for knowledge provides a re-entry into the cycle with new problems.

Organization of Topics

Preparation for Data Collection

Making the decisions that are necessary, prior to collecting field data and expressing those decisions in a research plan, is an iterative process. Ideally the researcher decides exactly what data are needed and how they will be obtained before going into the field, but this ideal is seldom fully achieved. Usually preliminary checks in the field reveal that initial expectations will not produce the kind of data needed to solve the research problem. When this is the case, a second attempt at specifying definitions and procedures is necessary. A trial run to evaluate the planned field techniques may indicate that the plans still are unsatisfactory. This back-and-forth process can conceivably be carried out several times.

Because these various preparatory decisions and activities are seldom accomplished without checking and re-doing some of them, there is no universally accepted order for their completion. To a degree, however, the next four chapters are arranged in the sequence that researchers usually consider them. Chapter 2 continues this discussion about the initial preparation for data collection. Chapter 3 deals with questions about how to measure the phenomena to be studied, and Chapter 4 presents some decisions and procedures involved with selecting the specific individuals from which data will be obtained.

Techniques for Collecting Data

Chapters 5 through 9 describe various techniques that are used to collect data for geographic problems. The techniques are classified into three broad types according to a level of "filtering," that is, to the various amounts of interference that may occur between the observer and the object being observed. At one level the person directly observes the phenomenon at its specific location on the earth; at more distant levels the researcher depends on an instrument and/or another person for the data. These three levels are described below with the aid of an illustrative diagram (fig. 1.4).

Direct Observations

Level 1 includes those techniques that require the physical presence of the researcher at the location of the observed phenomenon. This class embraces many of the traditional geographic field techniques that "get one's boots muddy" when tramping over the countryside looking at the natural (N) and cultural (C) features. It also includes field methods used to gather data about activities of people (P), which can be observed directly (Po). In general, Level 1 techniques are those dependent on the field worker receiving stimuli visually, audibly, or through one of the other human senses, which means that invariably the researcher must go to where the phenomena are located to see or hear about the aspects selected for observation.

Chapter 5 pertains primarily to visual observations, but certainly effective researchers are sensitive to all stimuli that may provide information. Several techniques combine listening with seeing. Conceivably a researcher could study olfactory variations (e.g., odors in a polluted region), or sense areal differences in phenomena by tasting and touching, but these human sensors are rarely used consciously. Chapter 6 is organized around techniques that measure attributes by obtaining respondents' expressions. These are often expressed orally but not to the exclusion of visual expressions. Although the researcher may personally hear the utterances of respondents (i.e., the oral expressions, symbolized as $P_{E,O}$) or see their written responses to questions (symbolized as $P_{E,W}$), additional filtering does occur. The generators of these stimuli are persons who themselves may restrict the flow of information about their feelings, beliefs, and similar intangible characteristics. These restrictions that are imposed, even unconsciously, by respondents produce additional filtering, which often affects the quality of collected data.

Even though the researcher may personally sense the objects of the real world, direct contact does not assure accurate data because the process of observing in itself involves a kind of filtering, thereby often introduces distortions. The observation process involves (1) receiving stimuli from the phenomenon, (2) mentally manipulating the impression of those stimuli in order to interpret them, (3) combining the impressions with other previous impressions and their interpretations,

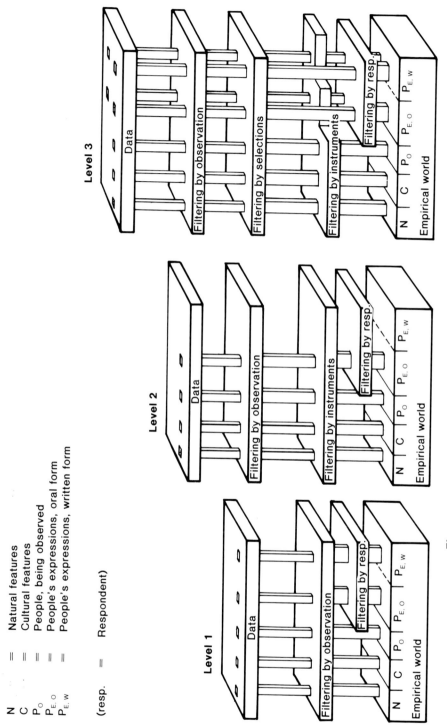

Figure 1.4. Filtering in the process of collecting and recording data

N = Natural features
C = Cultural features
P_O = People, being observed
$P_{E,O}$ = People's expressions, oral form
$P_{E,W}$ = People's expressions, written form

(resp. = Respondent)

and (4) deducing an interpretation of the phenomenon. This process is a type of filtering that produces unreliable or inconsistent data by different observers and by the same observer under varying conditions. Optical illusions, emotional distortions, and cultural preconceptions are some of the human frailties that may affect observations. Experimental psychology has demonstrated that humans selectively see what they want or expect to see. The results of inconsistent human observations are illustrated by the contradictory statements by witnesses who observe an accident or similar catastrophe, by contrasting reports from those who have or have not seen a UFO, by debates over the authenticity of sightings of the Himalayan "yeti," and by disputes over the calls by sports referees. Certainly there are too many cases where distortions have obviously occurred to accept wholly the declaration that "seeing is believing."

Persons who collect field data must be constantly aware of the human filters that distort observations of the empirical world. The realization that each person employs a unique selectivity when observing phenonena impels some scholars to reject the methods of scientific positivism because the goals of perfect objectivity are unattainable. A major emphasis in this book is that personal biases and subliminal selectivity do exist, and consequently every field worker must adhere to procedures designed to minimize the distortions inherent in collecting data. In spite of its limitations, however, the scientific methodology does provide a valuable framework for guiding researchers in seeking answers about locational problems.

A second form of filtering occurs when observed impressions and sounds are converted into numbers, words, and other symbols. The data that are actually manipulated and analyzed are only symbolic representations or abstractions of the actual phenomena. Ideally the isomorphism between the abstract symbols and the real world objects is unambiguous. In actual practice, however, representations are often only rough approximations. These, in turn, may be misinterpreted and omit essential attributes.

The two filtering sources can be illustrated by describing some facts observed in Choros. Personalized selective filtering occurs when two individuals look at the existing school building. Dr. A. observes an old, outdated structure located in a deteriorating part of the city. To her it is clearly an unattractive building standing in an undesirable locality. Mr. C. sees a school building that is—and has been all his life—a familiar landmark within his home neighborhood. He views it as an attractive feature because it is a positive element in his "sense of place." Clearly the same structure is viewed differently through the perceptual filters of Dr. A. and Mr. C.

The second filtering (i.e., that by abstraction) occurs when Ms. S. examines a map of Choros which includes the location of the elementary school building by use of a map symbol. By studying the map and the building's symbol, she can extract and use various kinds of locational information. However, this cartographic representation provides only very limited information. For instance, it does not provide data about the adequacy of fire doors in the building. A second Choros illustration involves information that was recorded in numbers. Dr. A., when gathering data about the present condition of the school building, measured the number of square feet of windows in the building and then recorded those measurements on a sheet of paper. These numbers convey the selected facts about window space, but they do not provide other facts (e.g., about how often the curtains are drawn or about the number of times students and staff glance outside during a typical day). Although she directly observed the windows in their entirety, Dr. A. selected and recorded only one characteristic. Her record of numbers does not contain all the attributes of the original phenomenon.

Selectivity is necessary for collecting, storing, and analyzing data. Almost by definition, data analysis requires the conversion of field observations into an abstract form. Researchers must accept the fact that abstraction is essential. With this acceptance goes the responsibility of always being aware that all field data are selected abstractions and that they should contain a minimum of distortion.

Data Sensed by Instruments

Level 2 consists of those field techniques that produce data through the use of instruments and their recorded output. The camera with its record in the form of a photograph is a common collector of visually-sensed data. This instrument is so commonly used by geographers that photogrammetry and air-photo interpretation are often studied separately from other field techniques.[6] In recent years the rapid expansion of other remote sensors of the electro-magnetic spectrum has led to special training courses for interpreting these additional remote images (see more in Chapter 7).

Instruments that record and preserve sound, such as tape recorders, are useful in collecting data which are expressed orally ($P_{F,O}$). Also, printed questionnaires might be conceptually considered as "instruments" that aid in recording and preserving written expressions ($P_{F,W}$). Omitted from the text are comments about specialized equipment designed to measure very specific attributes of phenomena and requiring technical training for their operation. Because of the emphasis on human phenomena, this exclusion is especially true for instruments devised for measuring features of the natural environment.

Using instruments to sense and record data mechanically has advantages (e.g., objectivity and speed in recording certain attributes); but they have the disadvantage of being yet another filter (fig. 1.4). One aspect of filtering is the possible lack of fidelity and the addition of "noise" that occurs in the instruments and their recordings. Another disadvantage is that in many respects they are not so versatile as humans. They are normally designed to sense and record only limited aspects of phenomena. The resultant selectivity is not entirely a disadvantage. In some situations this objective filtering aids the researcher in making purposeful selections. Nevertheless, the dependence on the selective output of instruments without any of the benefits of direct observations usually limits the quality of data. Without having actually observed the phenomena being recorded by the instruments, the researcher is deprived of the advantages of direct observation and the recall from the existential encounter. One who uses data recorded under unknown conditions by unfamiliar instruments operated by unspecified persons must rely upon data that were transmitted through several filters with their varying possibilities for distortion.

These comments about the limitations of instruments are not intended to deny the tremendous importance of data gathering techniques that produce photographs, tape recordings, and other records. The warnings are to remind the researcher of the potential instrumental distortions, which should be included in any assessment of alternative techniques for gathering data.

Stored Data

The researcher who examines and selects information from stored materials, (that is, data which have been acquired and recorded by someone else) is using a technique that belongs to Level 3. The data sources may be diaries, films, newspapers, taped speeches, census figures, maps, and other archival materials that are stored and are available to researchers. What is common to the techniques classified at this level is the filtering that results from observational and storage decisions by persons other than the researcher. Because one or more persons made decisions about which materials to observe, record, and subsequently save, the available data have been screened by personal, and usually unknown, factors. This prior activity results in a minimum of two, but often more, filters (fig. 1.4). If written materials (e.g., an old diary or other $P_{E,W}$ phenomena) are not regarded as "instrumental" records, then the two filters consist of (1) the selectivity of the writer in perceiving and recording observations about the location of phenomena and (2) the researcher's intrinsic observational interpretations (see Level 1). Also, when an interviewer asks an informant to provide information about phenomena other than the respondent's own self, this Level 3 technique may be regarded as having only two filters. For inanimate phenomena (N and C features) as well as external observations of persons (Po), the techniques of Level 3 involve three filters: those of recording instruments and the stored records, those of the persons selecting the data for recording and storage, and the researcher's own unique perception of the stored materials.

Often researchers using stored materials rely on data that have been filtered numerous times, such as illustrated by census data. After the governmental field workers obtain answers to the census schedules, the results are screened, organized, tabulated, and then published. In this case the eventual researcher who uses some of these figures for research does not even observe the field notes; thus, a large amount of filtering exists between the original phenomenon observed by the census-takers and the data selected for final use. Likewise, a map, which is an extremely valuable source of geographic data, is the result of filtering by the original surveyors and their instruments and by the cartographer who selected and symbolized the field data. Therefore, techniques that produce data by selecting information from a map generate highly filtered data.

It might seem that the disadvantages associated with greatly filtered data would diminish the utility of Level 3 techniques. This is generally true when the more direct methods of observation are feasible alternatives. However, some measurements, especially of historic phenomena, cannot be made directly; so the only option is to use stored data. When the researcher acknowledges the limitations, these techniques (discussed in Chapter 8) may aid in acquiring suitable data for solving geographic problems.

Not all archival materials are considered in Chapter 8 because concern is with the extraction of "new" data that contribute to the solution of geographic problems. The conceptual difference between using stored materials as a primary source for "new" data and assembling them as secondary materials is not sharply defined here. In general, techniques associated with "library research" that consist mainly of reorganizing data into a term paper or report are excluded.

Stored data that do not include locational information are also excluded from this text. For instance, a scholar seeking some answers to a geographic problem pertaining to a period fifty years earlier may initially view a long-neglected pile of old letters in an attic as a great treasure of potential data. If, however, after many hours of reading without finding any clues to the locations of the writers or their environments, the disappointed seeker may regard the letters as only dusty, mouse-chewed rubbish.

Summary

Field techniques can be classified into three levels according to the amount of filtering that occurs between the researcher and the phenomenon being measured. The classification of techniques, besides giving a rationale for the organization of chapters, provides a comparative ranking of techniques on the basis of their filtering. As a general rule, techniques that involve less filtering can be expected to yield more reliable data, when other factors affecting data reliability are held constant.

The merits of the classification scheme should not be overrated. Like many classification schemes, this one is plagued by types that are transitional and fit more than one category. Some transitional types are mentioned in this introductory chapter (e.g., respondents who provide information about themselves as well as serving as observers of other phenomena); others appear in later chapters. Also, the classification scheme does not necessarily guide the amateur researcher to the single most appropriate procedure because sometimes more than one technique is beneficial. By using multiple techniques (which sociologists call triangulation), the researcher may take advantage of the strengths of each strategy and can double-check the accuracy of the data. For example, a researcher might classify the economic use of an area by direct observations of the land (Level 1) and also by interpreting the patterns displayed in aerial photographs (Level 2). Another person might interview a group of informants (Level 1) and tape their conversations (Level 2) for later replay and analysis. These combinations of techniques that cross classificatory lines (levels) tend to prevent any clear-cut ranking of techniques according to the degree of filtering.

In spite of classificatory difficulties, the organizational framework should aid the reader in understanding the sequences for the presentation of techniques discussed in this text. Furthermore, the categorization of each technique according to its associated level of filtering may provide a preliminary aid to the reader who is considering the relative advantages and disadvantages of various options in collecting data. Further examination of the merits and limitations of each field technique constitutes the core of Chapters 5 through 9.

Evaluation and Applications

This first chapter does not deal with any particular techniques for gathering data; thus specific techniques are not evaluated at the end of this chapter as in other chapters. Nevertheless, the importance of having a goal for collecting data is stressed, primarily in terms of having a geographic problem that needs solving. Subsequent chapters assume the reader has a research problem to solve and needs a specific field technique that will provide the most effective data.

For readers who wish to pursue the ideas introduced in this chapter, the *Notes* suggest some pertinent references. Also, the following tasks may initiate some thoughts about how the contents of the chapter can be applied to other situations.

1. Describe how you could translate each of these phenomena/concepts into an *operational definition:*
   Grain farmers
   Shoppers in a city
   Geography textbooks
2. Write a *statement of problem* for each phenomenon below (a) worded so it is a geographic problem, and then (b) worded so it is not a geographic problem:
   International oil consumption
   Waste from nuclear plants

Notes

1. A person unfamiliar with contemporary geography might consult the following article and its bibliographic references as an easy introduction to the discipline: M. K. Ridd, "On Geography," in G. A. Manson and M. K. Ridd, eds. *New Perspectives On Geographic Education: Putting Theory into Practice* (Dubuque, Iowa: Kendall/Hunt, 1977), pp. 13–38.

2. The term "principle" overstates the kind of relationships usually established in a single empirical study. The word is used in this text as a shorthand for a "generalization based of a relationship observed between two variables."

3. The term "field work" has a wide variety of meanings. This is evidenced partly by the variation in courses taught as field geography in American colleges. These differences are described by the Commission on College Geography, *Field Training in Geography*, Technical Paper No. 1 (Washington, D.C.: Association of American Geographers, 1968); J. C. Friberg, *Field Techniques and the Training of the American Geographer*, Discussion Paper Series, No. 5 (Syracuse, N.Y.: Department of Geography, 1975); J. F. Lounsbury and F. T. Aldrich, *Introduction to Geograpic Field Methods and Techniques* (Columbus, Ohio: Charles E. Merrill, 1979); K. F. Nordstrom, "The Field Course in Geography: A Conceptual Framework," *Journal of Geography*, 78 (1979), pp. 267–272; and S. M. Thomas, "Some Notes on the Status and Nature of Field Method Courses at Colleges and Universities in the United States and Canada," *The Professional Geographer*, 30 (1978), pp. 407–412. In general, the publications prepared for teaching geography through field activities at intermediate levels of schooling, which have limited value for the topics in this text, are typified by J. E. Archer and T. H. Dalton, *Fieldwork in Geography* (London: B. T. Batsford, 1968); T. Bolton and P. A. Newbury, *Geography Through Fieldwork* (London: Blandford Press, n.d.); A. Gunn, *Techniques in Field Geography* (Vancouver: Copp Clark, 1962); R. N. Thomas et al., "Teaching Geography Through Field and Archival Methods," in G. A. Manson and M. K. Ridd, eds., *op. cit.,* pp. 263–289; and K. S. Wheeler and M. Harding, *Geographical Field Work: A Handbook* (London: Blond, 1965). Some of the traditional materials on field work tend to cover only a few techniques: C. M. Davis, "Field Techniques," in P. E. James and C. F. Jones, eds. *American Geography: Inventory & Prospect* (Syracuse, N.Y.: Syracuse University Press, 1954), pp. 496–529; R. S. Platt, *Field Study in American Geography: The Development of Theory and Method Exemplified by Selections,* Department of Geography Research Paper No. 61 (Chicago: University of Chicago Press, 1959); and R. S. Platt, "The Role of Field Work," in F. E. Dohrs and L. Sommers, eds., *Introduction to Geography: Selected Readings* (New York: Thomas Y. Crowell, 1967), pp. 183–187. Several texts that deal with "field research in the social sciences" tend to focus more on sociology and anthropology than on human geography; see the Notes for Chapter 5. References more applicable to the contents of this text as a whole are: R. Daugherty, *Science in Geography: Data Collection* (London: Oxford University Press, 1974); J. C. Friberg, *Fieldwork Techniques: A Revised Research Bibliography for the Fieldworker and Reference Guide for Classroom Studies,* Discussion Paper Series, No. 14 (Syracuse, N.Y.: Department of Geography, 1976); J. F. Lounsbury and F. T. Aldrich, *op. cit.;* and especially A. V. T. Whyte, *Guidelines for Field Studies in Environmental Perception,* MAP Technical Notes 5 (Paris: UNESCO, 1977).

4. The philosophical base for this text is frankly logical empiricism (positivism). For contemporary discussions about alternative philosophies, see Note #12 in Chapter 5 plus the following: J. N. Entrikin, "Contemporary Humanism in Geography," *Annals, Association of American Geographers,* 66 (1976), pp. 615–632; M. R. Hill, "Positivism: A 'Hidden' Philosophy in Geography," in M. Harvey and B. Holley, eds., *Themes in Geographic Thought* (London: Croom Helm, 1981, pp. 38–60; L. Guelke, "An Idealist Alternative in Human Geography," *Annals, Association of American Geographers,* 64 (1974), pp. 193–101; L. J. King, "Alternatives to a Positive Economic Geography," *Annals, Association of American Geographers,* 66 (1976), pp. 293–308; C. Mercer and J. M. Powell, *Phenomenology and Related Non-Positivistic Viewpoints in the Social Sciences* (Melbourne: Department of Geography, Monash University, 1972); E. C. Relph, "Phenomenology," in M. Harvey and B. Holly, eds., *op. cit.,* pp. 99–114; Y. Tuan, "Space and Place: Humanistic Perspectives," in C. Board, et al., eds., *Progress in Geography,* 6 (1975), pp. 211–252; D. J. Walmsley, "Positivism and Phenomenology in Human Geography," *Canadian Geographer,* 18 (1974), pp. 95–107.

5. For a more thorough discussion see: R. Abler, J. S. Adams, and P. Gould, *Spatial Organization: The Geographer's View of the World* (Englewood Cliffs, N.J.: Prentice-Hall, 1971); D. Amedeo and R. G. Golledge, *An Introduction to Scientific Reasoning in Geography* (New York: John Wiley & Sons, 1975); D. Harvey, *Explanation in Geography* (New York: St. Martin's Press, 1969); D. Johnson, "Mysterious Craters of the Carolina Coast," in S. Rapport and H. Wright, eds., *Science: Method and Meaning* (New York: New York University Press, 1963), pp. 64–86; T. S. Kuhn, *The Structure of Scientific Revolutions* (Chicago: University of Chicago Press, 1962); C. Lastrucci, *The Scientific Approach: Basic Principles of the Scientific Method* (Cambridge, Mass.: Schenkman, 1967); and D. Nachmias and C. Nachmias, *Research Methods in the Social Sciences* (New York: St. Martin's Press, 1976).

6. See Notes #2 and #3 of Chapter 7.

Chapter 2

Preparing for Data Collection

The previous chapter described the characteristics of geographic research problems and the ways in which the collection of data contribute to solving those problems. After the identification of a problem, such as locating a school building in Choros, the researcher must consider the steps necessary for solving it. The following subsidiary questions are discussed in this chapter:

How can a researcher obtain background information about a problem?
What decisions need to be made about operational definitions, the study area, base maps, and locational measurements?
What are the goals of a field reconnaissance, and how can they be achieved?
What can be accomplished by a pilot project?
How can large-scale field measurements be made cheaply and rapidly?

Purposes for Preparatory Procedures

An amateur researcher may be eager to plunge into the task of collecting data as soon as a research problem has been stated. However, several decisions must be made before truly meaningful data can be gathered. If these decisions are not made before going into the field, the researcher is apt to acquire faulty data and/or waste time with numerous false starts. The purposes for making preparations prior to the collection of the research data are (1) to increase background knowledge, (2) to make decisions about aspects of the research plan, and (3) to commence field arrangements.

Increasing Background Knowledge

Frequently a person realizes that a knowledge gap exists before completely understanding many of the factors that have produced a particular problem. When attempting to develop principles of location (by a Type II approach), a researcher needs to read the existing body of literature about the topic to comprehend what remains to be studied. For an applied problem (i.e., a Type III approach), the researcher must become familiar with the known principles that pertain to the specific conditions. From both approaches, therefore, the researcher must learn as much as possible about what is already known on the topic in general and about the locational principles of the specific phenomenon and related variables. It is always possible that such a search for pertinent information will uncover a previous study that essentially solves the problem and thus eliminates, or greatly simplifies, the task of collecting new data.

Also, the researcher should obtain background knowledge about the specific location to be studied. Gaining knowledge about the many distinctive characteristics of the study area is done best by going to the locality for an on-site examination. Direct contact with a region, which is a prerequisite to what is sometimes called "getting the feel of the region," is valuable because familiarity with an area can help in making realistic plans for gathering data. A person can attempt to learn about the human activities of an unfamiliar place by reading, watching TV, talking with friends, and examining maps, but these sources of information are often inadequate for discovering various nuances.

In summary, there are two primary objectives for increasing one's knowledge about the research problem: (1) to understand the general locational principles pertaining to the research phenomenon, and (2) to become familiar with the specific setting in which these principles are to be studied. A benefit associated with these two objectives, which frequently results from gaining more information, is further insight into ways to improve hypotheses. As mentioned in Chapter 1, more precisely stated hypotheses provide better guides about which data gathering techniques are most appropriate for solving the research problem.

Making Decisions for the Research Plan

Various definitions, procedures, and similar details must be clarified before a researcher can collect data that are relevant. These decisions may be incorporated into a formal proposal or research plan which may or may not be set forth in writing. Nevertheless, when researchers commence to collect data they do make numerous decisions, consciously or unconsciously, which are inescapable. The more effective procedure, therefore, is to confront these issues during the preparatory period of research.

These decisions do not necessarily fall into neat intellectual compartments, but six categories of decisions are suggested here. They involve the following: (1) operational definitions and individual geographic units, (2) a definition of the study area, (3) a base map, (4) a locational system, (5) measurements of field phenomena, and (6) a sampling design.

Operational Definitions and Individual Geographic Units

Most naming words are applicable to more than one member of a set. These general words should be deliberately translated into operational definitions that guide researchers in the field because the identification of individual members of the general set is inescapable. Whenever researchers measure specific characteristics of phenomena, they have adopted, consciously or unconsciously, an operational definition because it is impossible to collect data without some kind of translating scheme, whether specified or not.

The operational definition must describe as precisely as possible how individual objects are to be identified as belonging to the general set. The preparatory period is a good time for exploring various operational definitions. No matter how precisely a term initially seems to define what is and what is not a member of the set, there are many "shades of gray" encountered in the field. For example, Ms. S. believed that the volume of vehicular traffic on Highway #1 during the hour before and after school might affect the travel patterns of Choros pupils. Therefore, she planned to obtain data about the number of vehicles traveling on Highway #10 for designated hours of selected days. She had anticipated no difficulty counting the number of cars, motorcycles, trucks, and buses that passed the point in Choros where she would stand along the highway. However, her preliminary decisions about measuring vehicular traffic were put in questions when during her second reconnaissance to the field she tried a test count. She was surprised to see tractors and bicyclists on the street. Should she count bicyclists as part of the "vehicular traffic"? She also noticed that two cars filled with teenagers circled around and around her enumeration point five times during her brief count. Should multiple trips by the same vehicle be counted separately? One car slowed down and then parked along the street at the exact position she was standing. Should she add this automobile to her tabulation of total vehicles passing the observation point? It is true that Ms. S. could have anticipated all these "exceptions" to her initial (pre-reconnaissance) definition of "vehicular traffic" and procedures for its measurement if she had stayed at the university and read about other traffic surveys and pondered about the conditions in Choros. Her experience is typical, however, because researchers are seldom able to achieve a perfect prediction about specific variations in the pertinent phenomena in a study area. Very view people can express unambiguous definitions of a set of phenomena without having had some prior experience with a variety of individual members.

As an alternative to making a special trip to Choros for checking on her definition of "vehicular traffic", Ms. S. could have waited until she was ready to collect data on vehicular movement. In this case, she should do some preliminary checking on her tentative definition, make necessary adjustments, throw away her exploratory pre-adjustment counts, and then commence collecting data after settling on a realistic operational definition. Although such a strategy may be necessary when the study area is located a long distance from the researcher's home, the better procedure, when it is financially feasible, is to decide on an operational definition of the relevant phenomena during the reconnaissance/preparatory period.

Another decision, which is necessary when studying geographically continuous phenomena, is the adoption of a meaningful minimum areal unit. To illustrate, this kind of decision is being considerd by Ms. S. in anticipation of gathering data on urban land use. She plans to classify each parcel of land according to its economic use, for example, commercial, residential, or recreational. The entire parcel will be assigned to one class, so she must decide on the smallest unit of land that constitutes an areal individual. Suppose she chooses a city block. This areal unit is easily recognized in cities with rectangular street patterns and its land usage varies enough within most urban places that spatial relationships can be measured. However, she must realize that some other researchers may believe that an area the size of a city block often contains too much internal variation for precise analysis. If instead she chooses a city lot, it may reduce the number of different uses per areal unit (for a given classification scheme), but it will increase the total number of areal individuals within the study area. An increase in the number of individual units requires more time to collect data, a major disadvantage in some projects. Furthermore, even in small areas, internal heterogeneity is not necessarily eliminated. Even a small unit like a residential lot might be regarded as an area containing a dwelling, a building that stores a car, an uncovered parking spot, play space, and an open area of well-manicured vegetation. Therefore, even the reduction of the size of the areal unit from city block to a lot does not guarantee homogeneity of its use.

No standard guideline exists for making decisions about minimum areal units. The decision is one that each researcher must make after considering the merits of areal units of various sizes. It is not too early to commence considering different sizes at the beginning of the preparatory stage. This will allow the researcher time to note the possible impact of various areal sizes on other research decisions.

In some situations, the decision about the size of the minimum areal unit may be postponed. These situations occur when the researcher decides to observe a spatially continuous phenomenon at selected points rather than by small areas. The contrast is illustrated by comparing the procedures normally used to map urban land use and noise patterns in a city. Land use tends to be fairly uniform over small areas and then change abruptly. For example, one small area may be classified as retailing and another adjoining one as parking. The two activities (uses) are very distinct and are normally separated physically by a wall. To measure the variations of land use, therefore, is most effectively accomplished by observing the characteristics of each minimum areal unit. In contrast, noise levels tend to change spatially in a more gradual manner. This kind of variation can be measured by sampling noise levels at representative positions (see Chapter 4). To observe continuous phenomena at points rather than within small areas means that a decision on the size of a minimum areal unit can be postponed until a later stage in the research project. However, the person analyzing the spatial variation in the phenomenon must ultimately decide on the extent of area to which each point applies (represents). This is evident when data are mapped by using isolines because the map-maker must decide how to interpolate values between the known control points.

Individual geographic units for phenomena that are spatially discrete do not involve a decision about a minimum areal unit. Instead, the primary decision is one of identification. Although this is basically another facet of operationalizing terms, it concerns measuring the phenomenon rather than its spatial position. It is included, therefore, with other aspects of measurements in Chapter 3.

Definition of the Study Area

A statement of the research problem normally specifies the area and time period to which the problem pertains. It might appear that such a specification would remove the need for any additional decision-making about the study area. Indeed, when the methodological approach involves applying general principles to a specific problem, the study area is effectively defined. Usually the definition of the study area is in terms that make it easy also to delineate the actual boundaries in the field.

Dr. A. and Mr. C. view the school building issue as a problem that applies specifically to the Choros school district. There is no uncertainty about the study area—it is the jurisdictional territory of the school district. Furthermore, as residents of the city, both Dr. A. and Mr. C. know from experience where the boundary lines exist and they can point out the locations of the lines as they travel around Choros.

The task of defining the study area is more complex when the goal is the development of locational principles. This approach (i.e., Type II) requires the researcher to use the data collected within a specific study area to generalize relationships which will then pertain to other areas. Generalizing results raises the question about the representativeness of the study area as an example (sample) of many areas that are similar. The factors that must be considered in making such a decision involve sampling (which are discussed in Chapter 4). However, the nature of the decisions can be illustrated here.

For her research on the locational relationships associated with schools in small cities, Ms. S. selected Choros as representative of the set of small cities. By doing so, Ms. S. laid the foundation for a more specific decision about the study area. At the more precise level she needs to identify the de facto boundaries of "this small city." Should the boundaries be defined in terms of the school district, the administrative jurisdiction of the city, the area of high density which includes the new housing development outside the present city limits, or the total trade area? When Ms. S. decides the answer to that question, she will have established the criteria for identifying the study area. The remaining task will be to delineate the area in the field. If, for example, she decides that a study area defined in terms of high density is the most representative, then she still will have to decide how to demarcate that area around Choros.

To summarize, defining the study area for research designed to detect locational relationships (by the Type II approach) demands answers to three basic questions: (1) What area typifies the relationships (expressed or implied) in the research problem? (2) What criteria will identify the extent (boundaries) of the study area? (3) How can the boundaries be detected and demarcated in the field?

Base Map

A map of one's study area is very helpful for identifying places and routes during an initial field reconnaissance, recording basic areal data, analyzing spatial relationships, and presenting the results of geographic research. Because of this heavy dependence on a map, the researcher must carefully select an appropriate map, or maps, for these several tasks, giving primary emphasis to the map that is used for recording data in the field. Although the same map (or duplicates of it) may be used in other stages of a research project, the one on which field data are placed is called a *base map*.

The base map should be selected according to its scale, areal size, and detail. The interrelationships among the size of the study area, a convenient map size, and scale place limits on the suitability of various maps. The limits vary from one whose scale is too small for adequate areal precision to one with a scale so large that the map itself is unwieldy for outdoor work.

Maps with very little locational information do not make appropriate base maps for recording data because they make it difficult to find the map-to-earth and earth-to-map correspondence. In constrast, a city map that includes all buildings with their ownership, their utility connections, and their assessed value plus all topographic and vegetational features might be too clustered or "busy" with extraneous data to serve as a suitable base map. However, because such maps undoubtedly contain several features which can be identified easily in the field and thus serve as locational control points, they can be used to produce a satisfactory base map. By darkening or otherwise accentuating selected features, subduing or eliminating extraneous markings, and then duplicating the modified version, a researcher can easily create a suitable map for field work.

The decision about an appropriate map depends, in a practical sense, on the particular maps that are available. In the United States one of the most common is the U.S. Geological Survey (USGS). Its maps are available for all areas of the country, often with a choice of scales (e.g., at 1:24,000, 1:62,500, and 1:250,000) and sometimes with an option of multiple editions having different dates and data. Although these maps always include topography, they also symbolize many cultural (CV) features, which can serve to identify locations of a base map. Other map sources are various U.S. government agencies (e.g., the Forest Service, the Soil Conservation Service, the Bureau of Reclamation), departments and offices of various states (e.g., highway, planning), city offices and organizations (e.g., planning and election offices and Chamber of Commerce), and private companies (e.g., Sanborn city maps, gasoline retailers). A comprehensive collection of maps from many of these same sources is available through the National Cartographic Information Center (NCIC), which has affiliated centers in most states. Also, map collections in large libraries are another potential source a researcher should explore when searching for and deciding on a suitable base map (see Chapter 8 and Appendix B, Table B.2).

Locational System

Geographic data are empirical facts about places; therefore, one of the tasks in designing a geographic research plan is to decide on a *locational system*. The system must be capable of determining the location of places in the field and indicating their corresponding position on a map.

One locational scheme may be termed a nominal one because it relies on the use of names for places. For example, two highways are named "#10" and "#37", and signs with these numbers are posted along the highways. When Ms. S. first arrived in Choros she found the spot on the earth where these two named highways intersect. Later she spotted a building with a sign "Choros City Library," so she could identify the earth location known by this name. After receiving suggestions from local residents about where she might get something to eat, she found the place called "Marie's Cafe". These illustrate how she depended on finding certain places on the earth by communicating with persons who associated a name (a nominal designation) with each of the unique positions.

For geographic research the nominal system of recording position is unsatisfactory by itself because spatial relationships are impossible to establish. This is apparent if one attempts to map places that are named but possess no geometric information. For example, on a piece of paper one might be able to draw symbols representing Marie's Cafe and other features identified by Ms. S., but not be able to place them in their true relative locations without having some locational measurements.

The nominal system is useful for geographic analysis only when a master map provides the complete correspondence between each named place and its respective position on the earth. If the Choros City Library and Marie's Cafe were included on a master map (say, on fig. 1.1), then Ms. S. could effectively communicate additional facts about these few positions in the city. Another illustration is a map that accurately represents the location of all the counties and their respective names in a region in which data are collected by county name (as for an agricultural census).

The systems that specify locations without dependence on a map are those that utilize (1) a distance and a direction, (2) two directions, and/or (3) two distances. By starting from a base point any position on the earth (and on a map) can be determined with the combined measurements of distance and direction. From the intersection of Highways #37 and #10, Ms. S. can find the exact location of the Choros elementary school building by traveling for a stated distance in a specified direction. If some buildings hinder a direct route from the intersection to the school, the position of the school could be specified by a series of oriented distances. Similarly, after arbitrarily selecting a point on a proto map, the correct location of the school building can be represented by designating a direction, α, and a distance, d (fig. 2.1). This locational system, using polar coordinates, is the easiest of the three systems to use in the field where no other locational guides exist (e.g., see comments below about orthogonal grids).

A third position can always be determined by the intersection of two rays that originate at the ends of a base line (fig. 2.2), except when the rays have the same orientation and so are

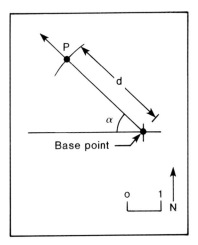

Figure 2.1. Locational system 1, a position determined by a distance (d) and a direction (α) from a base point

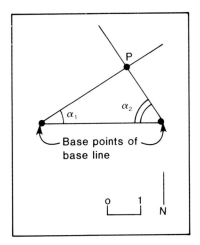

Figure 2.2. Locational system 2, a position determined by two directions (α_1 and α_2) from a base line

parallel. Therefore, after two points, either on the earth or on a map, are established, a third point can be indicated by specifying its directional positions, α_1 and α_2, from those two base points. The primary advantage of this system is the ease by which earth positions can be mapped by geometric triangulation.

The most common locational system measures two distances, d_1 and d_2, each from one or two orthogonal base lines (fig. 2.3). Most persons who read maps expect to locate places by using a pair of values associated with a rectangular grid. This system is illustrated by latitude and longitude values and also by the numbers or letters along the edges of a road map. For field locations the merits of this system depend on the manifestations of a grid network in the landscape. It is extremely rare that a person ever knows a precise position by its latitude-longitude intersection. Thus, this particular world-wide spherical grid system is not very helpful for most field work in local areas. However, where roads follow section lines of the U.S. Land Survey System or where streets form square blocks it is simple to specify a position by the coordinate values of an orthogonal grid. Where such a system is not visually obvious, the task of measuring distances from a pair of perpendicular base lines is not easily accomplished in the field.

The relative merits of these locational systems can be evaluated with reference to the equipment available for field measurements (see below), schemes that are already expressed in the field (especially rectangular grids), the detail on existing maps, and the desired degree of areal precision. There is no compelling reason for restricting measurements to only one system; thus, the researcher may decide to combine two or all three locational systems. What is critical is the correct use of a scheme that translates observed earth locations to a corresponding set of geographic facts. This declaration may seem obvious, but it emphasizes the distinctive nature of geographic facts. Since a geographic fact is composed of both a location and a phenomenon, the task of collecting geographic data requires accurate observations of both components. The locational component may seem easy enough from the perspective of an armchair traveler or researcher.

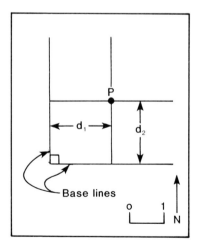

Figure 2.3. Locational system 3, a position determined by two distances (d_1 and d_2) from two orthogonal base lines

Achieving accuracy of location in the field, however, is sometimes more difficult, as evidenced by the sport of orienteering (i.e., depending on a map and compass to race across an area of rough terrain) and as testified by anyone who has been thoroughly lost while holding a map in hand.

Measurements of Field Phenomena and Sampling Design

In addition to decisions about operational definitions, the study area, a base map, and a locational system, choices must be made about ways to measure phenomena and about procedures for sampling. There is a wide variety of techniques that are employed for measuring a diversity of phenomena; therefore, the factors that affect this decision are by-passed here but considered in Chapter 3. Likewise, issues associated with sampling decisions are complex enough that they too are discussed separately (Chapter 4). Even though these decisions are discussed as separate issues, the reader should remember that they are highly interrelated and need to be decided almost concurrently with the other questions associated with the preparatory period of research.

Commencing Field Arrangements

Besides increasing the researcher's background knowledge and considering decisions about a general research plan, a third purpose for making preparations prior to actual data collection is to set up field arrangements. These arrangements are primarily for enlisting local cooperation. Although various undercover groups may penetrate a community and gather facts without the conscious cooperation from members of that community, this is not the normal situation for most research. In fact, most research is usually undertaken for the benefit of a community, either directly or indirectly through long-range societal gains, and cooperation by members of the community is essential for success in achieving that goal.

Cooperation from a community normally is gained quickest by contacting persons with power and influence. The relative importance of such contacts will vary with the organization of the community, the phenomena being studied, and the techniques to be used in collecting data. In any case, the researcher should attempt to establish contacts early in the preparatory period. Finding persons with power and influence who are willing to aid in a project is not always accomplished quickly.

Special permission is often needed to carry on research activities. For instance, when data collection depends on access to records and stored materials, the researcher should verify during the preparatory period that persons controlling such materials will allow them to be examined. Likewise, when the technique involves human beings who are under the authority of someone else (e.g., minors, prisoners, elderly persons in special homes), prior permission is needed to interview them. The prudent researcher will ascertain at an early stage whether the prospective subjects will be allowed to provide the desired data.

Preliminary contacts may not always need to be made with recognized community leaders. However when one's research involves groups of persons somewhat outside the mainstream of respectable, prestigious, and/or legal activities, procedures will usually differ from the norm. Gaining entfee to such groups is often enhanced by attaining the cooperation of a few influential members ("gatekeepers"), a task that may challenge the skills of an amateur researcher.

The preparatory period is also a time to arrange for equipment (e.g., a vehicle) and/or assistants. If research is conducted in an area foreign to the researcher, translators or other field assistants may be essential.

Techniques

Increasing one's background knowledge, making critical research decisions, and making field arrangements should be accomplished prior to collecting data. Techniques that apply particularly to these preparatory activities include a review of the literature, a field reconnaissance, a pilot project, and locational measurements.

Review of the Literature

Searching a library for publications that report on previous research can accomplish several goals. The existing literature may provide background knowledge about general principles of location (i.e., the geography of the research phenomenon). It may describe the specific study area (information which is especially helpful for research in foreign areas). And, it may suggest ways previous researchers have collected data for similar problems.

This text does not focus on techniques for conducting a library search for secondary materials per se, but it does deal with using archival materials and stored documents as sources for collecting new data. Comments about finding such stored materials (Chapter 8) may provide the reader with hints about pursuing library searches.[1]

Field Reconnaissance

An important technique for becoming familiar with a specific study area is a *field reconnaissance*. It is a preliminary examination by direct observation of phenomena in the specific area which is expected to be selected as the study area for the purpose of gaining an overall perspective. One objective is to acquire an understanding which supplements knowledge obtained from the literature review. If financial resources for the research project are limited, exploration during the reconnaissance should be done quickly and cheaply so resources are sufficient for the actual collection of data.

Traveling through the proposed study area is a common way for reconnoitering visual phenomena. A trip by car, train, or low-flying plane may provide an opportunity to view landforms, crop types, house styles, and similar landscape features.[2] A windshield tour by car with a few stops for closer examination is a familiar technique because such car "traverses" allow very rapid coverage of a large area, and they can even provide opportunities to take crude samples of continuous phenomena. For example, by making visual observations at regularly-spaced intervals (for instance, every fifth mile), a researcher could obtain a rough estimate of land use along the route. Such an estimate should be regarded, like other observations made during an initial reconnaissance, as only a first impression. The information may be biased (Chapter 4), and it should not be considered part of a subsequently acquired data set.

Sketching visually observed features is another way that some researchers gain preliminary insights into spatial relationships within the area. One of the benefits of sketching is that it forces the observer to select what appears to be most important in the landscape. Experienced observers know how to discriminate pertinent phenomena from trivial phenomena. Sketching may aid in developing a sensitivity for seeing and selecting meaningful relationships among phenomena.

Much can be learned about an area by talking with local residents. Visiting can take place through CBs, in taverns, at sports events, and other places where people are ready to talk to strangers. Conversations with public officials can usually yield multiple benefits: (1) The exchange establishes an initial contact that may be useful for subsequent arrangements and permission; (2) the officials usually possess considerable knowledge about the community; and (3) they may suggest other persons to contact.

Keeping a diary or a log of notes during field reconnaissance is usually worthwhile. It has the obvious advantage of being a potential record of names, places and facts that are obtained while scouting the study area. Also, it may serve as a record of tentative choices for definitions and similar decisions, and, therefore, as a skeleton of a research plan. Furthermore, keeping a log accomplishes some of the same goals as sketching, that is, it develops sensitivity to meaningful relationships without being restricted to visual features. Finally, a field diary in which a researcher records subjective impressions and reactions can aid the researcher in assessing those personal changes in perspective that may have affected the objectivity in collecting data.

Field reconnaissance refers to a wide variety of activities designed to help the researcher learn about the study area and to determine ways data can subsequently be collected. It tends to merge into a pilot study (see below), and it may be prolonged over a period of time if the benefits of greater familiarity outweigh the costs of time and travel. Some of the experiences of Ms. S. (excluding the time she got a traffic ticket because she became absorbed in looking around and consequently missed seeing a stopsign) may convey a sense of how she used a reconnaissance. When she first went to Choros she drove around the city to see the school building, features that indicate the economic base of a city, the quality of residences, and various movement of vehicular and pedestrian traffic. She noted, among other things, that when elementary pupils left their school building in the afternoon they tended to walk homeward in groups rather than singly by the most direct route. She decided she should sample some of the pupils' paths to see how their behavior might alter some of her initial assumptions about the proximity of school facilities to home residences. However, since the driving distance to the field was not great and because she had to hurry back to a campus meeting, she did not pursue this data source during the first reconnaissance trip. During her next trip she contacted Dr. A., who told Ms. S. something about the pedestrian habits of the children and gave her the name of the person heading the P. T. A. During the third trip, Ms. S. visited with the head of the P. T. A. about the possibility of working with parents who supervise children where they cross Highway #10. In fact, Ms. S. went to the field four different times to learn more about the Choros situation, to make contacts with various officials, and to try various data gathering techniques. In her case, then, the relatively low cost of a field reconnaissance, even though it involved four trips, was a wise use of her resources because it reduced the likelihood of what could have been several false starts during the more costly and critical stage of data collection.

Pilot Study

One of the purposes of the field reconnaissance is to explore various operational definitions and other options that are part of a research plan. Before making a final decision on these options, the researcher should make a trial run to see how well they function. The value of trying out operational definitions is illustrated by the experience of Ms. S., who thought she had a satisfactory definition of vehicular traffic in Choros until she actually did some counting and discovered gaps and vagueness in her initial definition.

A similar discovery of weakness in measurements can occur during a trial run. While in an office or armchair at a library, a researcher may develop what seems to be a very sensible method for measuring selected phenomena. Attempts at using the method in the field, however, may expose numerous difficulties and ambiguities. Preliminary tries in the field may reveal these weaknesses and thus allow the researcher to make revisions before investing time and effort in collecting the bona fide data.

A trial run of all the data-collecting procedures tentatively designed by the researcher is often termed a *pilot study* (or pilot project). It functions like the dress rehearsal of a theatre performance because its goal is to approximate the "real" performance while checking for any remaining flaws during this last "practice". The pilot project is the last chance, before collecting the field data, to verify that the concepts stated in a research problem can be operationalized effectively through specific field definitions, procedures, and measurement techniques.

The techniques associated with a pilot study vary as much as field techniques in general, so a discussion of their complete range would be equivalent to a review of this book. However, because the map is basic to all locational studies, it can typify the checking done during a pilot study. Assume a base map has been selected during the preparatory period. The task during the "dress rehearsal" is to verify that the map symbols do correspond to observable field phenomena and that relevant field features are represented on the map. The researcher should check to see that, in addition to a correspondence in the existence of features on the map with those in the field, there is accuracy in the locations of those features. Usually this involves viewing selected phenomena on the earth while simultaneously holding a map and looking at its recorded representations of the same selected phenomena. The most difficult aspect of this procedure when done only visually is verifying the locational accuracy of the map. More precise measurements than those achieved by "eyeball" estimates are discussed below.

Locational Measurements

A system for determining location is essential for data that are geographic. It is required for measurements in the field as well as for recording data on a map. During the preparatory stage a decision should be made about the system (one, or a combination, of those explained above) most appropriate for the study area and phenomena. Then the question arises: What techniques can be best used to determine locations in the field? An answer is given here under four headings: (1) Using Place Names, (2) Estimating Positions Visually, (3) Pacing Combined With Using A Compass Or A Transit, and (4) Triangulating With A Plane Table.

Using Place Names

The exclusive use of a nominal system depends on the existence of a map with the names of all the places pertaining to the phenomena being studied. The primary technique for its utilization in collecting geographic data requires finding the corresponding locations in the field, and by looking for signboards and asking persons about the names of places.

Locational data obtained from informants, either by listening to oral responses or by reading written materials and signs, are usually quite accurate for administrative areas, especially large ones such as cities, counties, and states. Although in the field it might be quite tedious to trace the entire set of boundaries for a study area, the determination of whether specific features and persons are or are not within a designated area is usually fairly easy. For example, Ms. S. can probably learn whether the "New Addition" is inside or outside the Choros city limits by merely asking a few residents.

For most studies that involve large areas where earth positions are adequately named and the referencing is moderately precise, the nominal system is quite practical. Maps already exist for most large regions, so usually there are no complications in obtaining a base map having a sufficient number of place names. It would be a waste of time and effort to measure positions more precisely when the research problem does not demand it.

When using historical materials as field data, the researcher may be entirely dependent on the nominal system because locations of various phenomena are often revealed only by names. With one or more master maps, a researcher can use data given by locational names to solve problems in historical geography.

Complications occur when the study area is small (e.g., a neighborhood, a section of farm land), lacks administrative definition (e.g., the Midwest, the inner city), or for any other reason is without a map equivalency. In fact, if no map exists, the nominal locational system cannot be used. Note that the absence of a map is a critical defect because a geographic problem cannot be solved without the kind of information that can ultimately be placed on a map. Geographers always require data that possess a one-to-one correspondence with earth locations and, hence, are mappable.

Sometimes a map with named locations exists for a particular area but is not directly available for field work because it is a rare copy and confined to its place of storage (e.g., library or office). The sizes of other maps may be unsuitable for carrying into the field. In these cases the research can (1) duplicate and, if necessary, change the scale of the existing map so it will be suitable for field work or (2) record the locational data by their names in the field and then later consult the master map.

Estimating Positions Visually

Another technique for determining positions in the field involves estimating relative locations by looking at (i.e., "eyeballing") identifiable features. This method is utilized primarily for large-scale studies where greater areal precision is required than is provided by administrative or other named areal units. Positions of the research phenomena are observed in the field by visually comparing their locations relative to identifiable landmarks and expressing those positions in descriptive terms. These geographic facts can be recorded in written form (e.g., a third of the distance along a line between the church and the school; directly across the highway from the delicatessen; approximately a half-mile southwest of the grain elevator) and later converted into their geometric positions on a map. Such conversion, by necessity, must include all the same landmarks or control points used in describing positions.

Most geographers find it easier and faster to mark locations directly on a base map than to describe the relative positions in words. Marking data on a map while in the field forces the researcher to estimate the position of each feature relative to other phenomena in the field and also to guess at the feature's corresponding position on the map. In other words, to place symbolized data on the base map accurately, a person must be able mentally to scale-down the estimated field positions to their corresponding map locations. Persons with basic map-reading skills can normally achieve enough areal precision through this visual technique to serve the locational requirements of most geographic problems.

Usually a researcher can find a printed map, which is suitable either as it is or with modification, to use as a base map for marking field data. If no satisfactory map exists, the researcher may be able to produce one by altering an aerial photograph. Conversion may be achieved by tracing or highlighting selected features on the photo and then duplicating the sketch on nonglossy paper. Through this procedure researchers can choose the features they want to identify in the field as locational control points. In addition to this flexibility in selecting control points, base maps made from aerial photographs may be superior to published maps because the former reveal the areal bounds of features rather than just the extent of a map symbol. Possible disadvantages, however, are the dependence on only phenomena visible from an aerial position and the existence of some locational distortion around the edges of most photographs.

Pacing Combined with the Use of a Compass or Transit

When very detailed geographic information about a small area is required, neither the place-name technique nor the visual estimation of locations is adequate. When these conditions exist, a researcher must consider alternative techniques that will provide greater areal precision.

The need for greater locational precision refers primarily to entering data on a base map, but in addition, there are times when a researcher needs to make an original map. If a large-scale map of the study area already exists, then the task is to add the locations of the field phenomena by map symbols. If a map (or aerial photograph) shows several features in the study area but its scale is too small to allow room for entering field data, then an enlargement plus possibly enhancement and other modification will produce a suitable base map. If no map can be modified to create a satisfactory base map, then the researcher may need to go to the study area and make an original map. In these circumstances, the researcher will need to utilize some of the fundamental techniques of *field surveying.*

Surveying the earth's surface can be accomplished with a high level of precision by using sophisticated equipment calibrated to make minute measurements.[3] Rarely does a geographic problem require locational data measured that precisely, and techniques requiring expensive instruments are excluded here. In contrast, the techniques described here are those that depend on very simple equipment but which utilize geometric principles to give greater locational precision than can be obtained by eyeballing.

As stated previously, the locational systems that depend on geometry require measurements of distance and direction. A fast, cheap, and reasonably satisfactory way of measuring distances is by *pacing.* This requires that the researcher establish the length of his/her average stride, which by standard definition is the distance between where one foot touches the ground in successive steps (but, if more natural to the person, the distance between the steps of both feet can be counted). To determine one's average stride length in terms of standard units, e.g., meters, the researcher should count the number of strides along a line a few hundred meters long over roughness conditions similar to those in the study area by going in each direction two times. The average (i.e., the mean) of these counts is considered the stride length. As an illustration, stride counts of 197, 204, 197, and 202 along a 300-meter stretch yield a mean of 200 strides for the total distance and a conversion factor of 1.50 meters per stride.

Measuring direction from an origin, say A, to another place, call it B, may be done simply by using either a compass or a transit. The *compass* is an instrument that indicates the direction of magnetic north in terms of angular units, usually degrees. If the compass operator stands at the base point, A, and points the compass toward an object at the next field position, B, the instrument indicates the angle between the ray AB and the magnetic north-south line through A. This angle may be expressed as the number of degrees clockwise from north, the azimuth, or the degrees east or west from north or south, the bearing. Even an inexpensive pocket compass is capable of detecting directions accurately enough for the data needs of most research conducted within areas no larger than a few square kilometers.

When using a compass a person may encounter errors in two ways which result from its magnetic properties. One way occurs because the compass needle points to the magnetic north rather than to the true polar north. This is not serious, though, because the compass orientation is virtually the same within a single study area; thus, (1) this constant factor can be ignored because relative positions will be accurate or (2) the magnetic declination from true north can be adjusted by adding or subtracting a constant angle to all readings. The other error, however, may be more serious. Certain metallic objects may affect the magnetism of the compass so it cannot be used to provide reliable measurements of direction when the study area contains many such magnetic disturbances. In this case a transit (see below) is the appropriate instrument for measuring angles.

Two procedures can be used to reduce errors and provide estimates about the severity of errors that do occur. One procedure is to repeat the measurements in reverse order. Strides that were counted by walking up a slope the first time are later counted while pacing downslope; angles are measured as different azimuths or based on different bearings. The other procedure is to establish the location of a position from more than one base (i.e., by performing a type of geometric triangulation). The nature and magnitude of discrepancies can warn a researcher about the severity of errors and the possible need for re-doing some measurements.

A *transit* is basically a sighting line mounted over an angular scale (e.g., a circular protractor) on a portable plane, normally mounted on a stand (fig. 2.4). The objective is to measure the angle between two rays, with the position of one having already been established. This is done by temporarily fixing the position of the working plane, which means it must be stationary during the period of the two measurements. First the direction of one ray is sighted and the angular position recorded. This is easily accomplished if the height of the stand holds the plane close to eye-level. Then the direction of the other ray is sighted (but without bumping the plane stand). The difference between the two readings provides the angle needed for mapping.

High precision transits with telescopic viewers are manufactured and can be used if the researcher has access to them. However, the basic components of a transit can be homemade with a line-of-sight pointer mounted over a smooth plane with protractor markings. Building a tripod to hold the plane may be a little more complex, but even something as common as a stool or short stepladder can serve as a stand where the ground is not too rough.

The assembling of locational data into an original map, or even the addition of supplemental positions to an existing base map, is achieved most successfully if the field data are well organized. The field measurements should be organized to indicate clearly the correspondence between distance and directional facts and their respective field positions. An amateur researcher should try

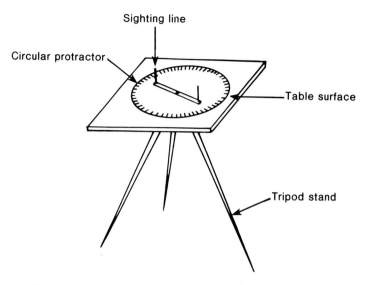

Figure 2.4. Simple transit for measuring horizontal angles

various formats to learn which organizational style is personally easiest and fastest to use in the field and is least confusing when later transforming the information into map positions. The format should be designed so the person while still in the field can check quickly for any omissions.

Triangulating with a Plane Table

A compass or transit can be used to construct a map that depends primarily on directional measurements taken from a pair of points (fig. 2.2). After the first pair of points are designated, usually as the end of a base line, any third position can be determined by measuring its angular direction from each of the "base points". Unless those two directional rays from the pair of base points are parallel, they always determine a triangle, which is the fundamental principle of geometric *triangulation.* Through repeated triangulation, subsequent field positions can be measured by their directions from any pair of known positions.

These field measurements provide the data for locating the isomorphic positions on a map. Mapping commences with a pair of points selected to correspond to the base points in the field. Subsequent points are located by reproducing the triangulation data.

An alternative procedure to collecting field data and later assembling them into a map is a technique that produces a map directly in the field. The procedure commences with a blank sheet of paper and ends with a map of field positions. The only equipment needed is an *alidade,* which is a sighting line affixed to a straight-edge, and a *plane table.* The plane table is a portable drawing board on which the proto map is secured at a comfortable height for sighting and drawing (fig. 2.5). Although high quality alidades with telescopic sights can be purchased, a person can construct a satisfactory instrument with a ruler and some slender nails. With this equipment, a map can be created by a series of triangulations that result from sightings taken at several field positions.

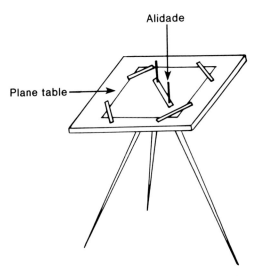

Figure 2.5. Plane table with simple alidade

On the blank sheet of paper the field worker should initially place two base points in relative positions that will produce an appropriate scale of map when they are matched with two field positions. The remaining activity consists of sighting toward features in the field while drawing corresponding lines on the map, an procedure frequently called plane-tabling (Appendix A).

The task can be used also to add locational data to an existing but incomplete map. When supplementing a map, the field mapper does not have the option of aribtrarily placing two base points on a blank piece of paper. The researcher must identify two specific positions on the map and their corresponding locations in the field. These two positions will represent the base points from which subsequent fields sightings and map rays will originate. The pre-existence of several map positions may reduce the number of triangulations required and provide more accurate checks for the places that are then added to the maps. It is essential, however, that the added positions are integrated correctly with the already established locations.

Evaluation and Applications

Evaluation

Several decisions about solving a research problem in a specific study area need to be made prior to the time of data collection. If that time is not until the researcher first arrives in the field expecting to gather data immediately, then costly blunders may occur. This happens because it is almost impossible to anticipate all the unique characteristics of phenomena within a particular locality. Therefore, the time and effort invested in making preliminary examinations during a reconnaissance and in trying out the potential techniques during a pilot study normally will reduce the total investment in data collection.

It must be recognized that in some cases a field reconnaissance may be too costly, or even impossible. The costs of travel to an area located a long distance from the researcher's home such as in a foreign country may deter a person from conducting a reconnaissance during the preparatory period. If the anticipated costs of a separate trip jeopardizes the entire project, the field reconnaissance phase may have to be sacrificed. In effect, the researcher must weigh the costs of time and travel spent in a reconnaissance and a pilot project against the expected benefits derived from advanced preparation. Weighing these costs and benefits is a subjective decision which each researcher makes, either consciously or subconsciously. A rational approach is to consciously consider the best way to prepare to collect data.

Applications

If you want to gain experience with some of the techniques described in this chapter try one or all of the following activities.

1. Participate in the next orienteering meet in your locality. The sport of orienteering is based on the ability to convert map information into field data and to measure distances and directions by pacing with a compass. Thus, orienteering should help you sharpen these skills.[4]

2. Make two maps of a small area, for example, a city park. For one, collect data on field positions by pacing and sighting with a compass, then assemble these data into a finished map. For the other, make the map in the field by using a plane table and simple alidade. Compare these two maps in terms of their accuracy and the time it took you to make each.

3. Take a "windshield" tour of a local area to gain familiarity with attributes of attractiveness. The need for evaluating scenic beauty, visual pollution, and similar esthetic intangibles has become quite important in recent years because of an increased concern about the ways people utilize their natural and manmade environments. To plan and manage natural resources wisely and to assess deterioration of manmade features require measurements of natural scenic beauty and urban structures. Appropriate techniques for measuring phenomena like "attractiveness" are discussed in Chapters 3 and 5, but at this time start considering several different kinds of phenomena (natural and cultural, intrinsic and situational) associated with attractiveness. Note variations in several phenomena within the selected area and ponder on how they might be measured. Try using a few criteria to see whether they might be suitable for measuring variations in visual attractiveness. Do not design a measurement scheme—just examine various possible criteria that you might subsequently use in a scheme.

Notes

1. For suggestions about using library materials, see R. W. Durrenberger, *Geographical Research and Writing* (New York: Thomas Y. Crowell, 1971); L. Haring and J. F. Lounsbury, *Introduction to Scientific Geographic Research,* 2nd ed. (Dubuque, Iowa: Wm. C. Brown Company Publishers, 1975).
2. C. C. Colby, "The Railway Traverse as an Aid in Reconnaissance," *Annals, Association of American Geographers,* 23 (1933), pp. 157–164.

3. Publications that provide more detail than here, but are not very technical are T. W. Birch, *Maps, Topographical and Statistical,* 2nd ed. (London: Oxford University Press, 1964); B. J. Garnier, *Practical Work in Geography* (London: Edward Arnold, 1963); J. G. Lounsbury and F. T. Aldrich, *Introduction to Geographic Field Methods and Techniques* (Columbus, Ohio: Charles E. Merrill, 1979); J. C. Pugh, *Surveying for Field Scientists* (London: Methuen, 1975).

4. W. P. Adams, "Geography and Orienteering," *Journal of Geography,* 71 (1972), pp. 473–480; B. Kjellstrom, *Be Expert with Map and Compass,* new rev. ed. (Harnsbury, Pa.: Stackpole Books, 1967); L. Carlson et al., *Introduction to Orienteering* (Toronto: Canadian Orienteering Service, 1971).

Chapter 3

Measuring Field Phenomena

This chapter continues the general topic of preparing for data collection by focusing on issues that pertain to measuring phenomena. Questions that are discussed specifically are the following:

What is meant by "measuring" phenomena?
What are the differences among nominal, ordinal, and parametric measurements?
What is meant by the validity, reliability, precision, and accuracy of data?
How can a person visually identify features in the field that belong to a particular definitional class?
What are the characteristics of a suitable scheme designed for classifying phenomena?
What are some of the ways of measuring non-visible attributes such as attitudes and feelings?
How can an index based on a questionnaire be developed?

Nature of Measurement

The collection of data refers to gathering information about properties of individuals (objects, persons, events, areas). The purpose is not to gather the individuals themselves like one would collect fossils or ferns; rather, the goal is to measure selected characteristics of individuals. *Measurement* is the procedure by which a researcher assigns numerals to objects, persons, or events according to prescribed rules.[1] Because the collection of field data is essentially the process of measuring phenomena, the implications of this definition of measurement need to be considered carefully.

In this definition the term "numeral" is regarded as a symbol that signifies identification or matching. A numeral should not be confused with the word "number" (see below). It is true that numerical digits may be assigned to objects, but these digits frequently do not have mathematical meaning. In many cases the numerals can be replaced with a name, a picture, or a different series of digits and still serve the same identification purpose. Examples are the numerals on shirts of athletes, the signs marking highways, and the set of digits that match a particular telephone with a unique circuitry. Also, an assigned numeral may refer to a class that differentiates an individual (object, person, event, area) from individuals in other classes. For instance, if all numerals between 20 and 39 were assigned to game participants who played a certain team position or if the first three digits of a seven-digit telephone "number" were assigned to phones located in a particular region, these numerals would indicate a measurement by class. When measurement of a characteristic is by matching it with a symbol or by identifying it with a class, the procedure yields results that are called *qualitative* data.

When a numeral is given a *quantitative* meaning, then it is properly called a number and can be used arithmetically. To extend our previous illustrations, the numeral on an athletic shirt could be called a number if it corresponded to the points scored by the wearer during the previous season; and the set of identifying digits for a telephone would become a number if it had a quantitative meaning, e.g., distance from the headquarters of the telephone company. When a researcher decides to measure an individual quantitatively, the operation is basically the same one-to-one matching procedure as assigning qualitative numerals but with the elements in the domain of real numbers. Thus, to measure one or more of the many characteristics that each individual possesses is to assign a quantitative number or a qualitative name or class.

Two sets of data that Ms. S. plans to consider may illustrate aspects of the measurement process. One phenomenon that she thinks might affect the desirability of a school site is noise level; therefore, she wants information about the spatial distribution of noise throughout Choros. Because this is an areally continuous phenomenon that changes through time she will need to select a few places at specific times to represent the total spatial and temporal variation. The selected places cannot be gathered up and hauled back to a laboratory—they must be observed in the field by measuring their associated properties. For each of these selected positions she must obtain data about the phenomenon of noise. This requires some kind of procedure that will react to the variations in noise and will assign a class or number to these variations. In this case she may decide to use a physical instrument (an audiometer) that is calibrated to provide numbers (decibels) equivalent to the volume of sound. Her main measurement task will be to operate the equipment correctly and to record the readings accurately, so the procedure of "assigning" values is fairly routine.

Since Ms. S. wants data that are geographic, she also must obtain information about the location of each site where noise data will be obtained. Regardless of the locational system she employs, her task in assigning numbers to this phenomenon, earth position, will consist mainly of utilizing instruments or counting strides.

The second set of data to be collected by Ms. S. requires data about residential locations and facts about attitudes towards the use of school buses in Choros. Measuring residential location involves almost the same procedures as measuring locations of sites where noise data are to be collected, although the locational system may differ because residences are discrete occurrences. What about the measurement of "attitudes toward school buses"? Remember that Ms. S. is not collecting individuals—she is collecting data that measures a specific property about each human individual. In this case the assignment of numbers or names to this property is less standardized because there is not a physical unit (like a kilometer) that is well known and communicated easily. She needs a way to measure this intangible characteristic of attitude such that each individual can be matched with a class or number. In this situation her procedure will probably be to construct a questionnaire (an "instrument") designed to measure attitudes about school buses. She will ask selected individuals to respond to these questions. She will then tabulate their answers by a predetermined manner which will produce an overall score for each person. By prescribing a specific procedure for matching scores with the sets of responses, she will be able to measure an attitudinal attribute of each individual.

In summary, measurement is the process of applying numbers or classes to the properties of individuals. Although laypersons may associate measurement with numbers, the definition presented here is more inclusive. Likewise, mathematical relationships do not require numbers because these relationships also include conditions of equality or belongingness. For example, to state that individuals X and Y both belong to the same class means that, at this level of measurement, X and Y are equivalent. Additional comments about various levels of measurement follow.

Levels of Measurement

Although there are varying thoughts among some scholars about the differentiation of measurement types, most researchers consider measurement as occurring at three or four levels: (1) nominal, (2) ordinal, and parametric, which is either (3) interval or (4) ratio.

Nominal

One kind of measurement involves determining whether an individual belongs to a named class or not. Determination is achieved by answering the question: To which of two or more classes does this individual belong? Is the respondent "male" or "female"? Is this land being used for "agriculture", "industry", "residences", "recreation", or "other"? Is this attitude "favorable" or "not favorable"? By assigning a name (or numeral) to each individual, the researcher measures phenomena at a level that is called *nominal*. As illustrated, the number of classes varies. If there are only two nominal classes, some persons may refer to the data as dichotomous or binary.

It appears that nominal measurement should be fairly easy because the data-gatherer needs only to determine the class in which the individual fits. If the research is classifying people by sex, it is generally easy to distinguish between males and females and assign each person to one of the two categories. Even if there are several classes in a scheme (for example, farm livestock classed as cattle, hogs, and sheep), making a decision about the correct class may not be difficult. More commonly, however, the researcher will encounter some bothersome uncertainties. One of these uncertainties pertains to the initial identification of the phenomenon itself. The other difficulty occurs when one must differentiate between classes. Both topics are discussed later in this chapter.

Ordinal

The next higher level of measurement includes all aspects of nominal data plus the ability to put classes in hierarchical order. When classes can be ordered, or ranked, the measurement is termed *ordinal*. If, for example, a person classified occupations only by name, no ordering is possible. If these same occupational classes were ranked according to "social prestige", then ordinal data would result. This illustration of ranking groups of persons by occupational class produces what is sometimes termed *weakly ordered* data.

If, in addition to being able to equate each individual with a ranked class, the researcher can rank all individuals within each class, then the data are termed *completely ordered*. In fact, when each individual is measured to the degree that he/she is greater than or less than all other individuals, a complete ordering is achieved and grouping by classes is unnecessary.

Interval

With an assignment of numbers, variation in a phenomenon can be measured at yet a higher level. At the *interval* level of measurement, differences between individuals are expressed by numbers that can be manipulated arithmetically. For example, if H, J, and K are ranked best, next best, and third best respectively according to some criterion, the produced data are completely ordered. If, however, the same three were measured by a scale that assigned the values of 7, 4, and 3 to H, J, and K respectively, the results are interval data. This level allows one to see that not only is H better than J, and J is better than K, but the interval between H and J is 3 units while the difference between J and K is only 1 unit.

To assign numbers to the properties of phenomena greatly increases the potential for statistical analysis. Because this is true, researchers sometimes make the mistake of attempting to measure phenomena by using interval data even when such a level is not supported by the instrument. This error is typified by the following situation. Ms. S. asked respondents to indicate their feelings about school buses by reading a series of statements and then marking each with one of the five categories: Strongly Agree (SA), Agree (A), Undecided (U), Disagree (D), and Strongly Disagree (SD). The results for four statements (I, II, III, and IV) by three respondents, P, Q, and R are given in Table 3.1 (Column a). Since respondents have measured their feelings by checking one of five classes, these results are normally considered ordinal data. If Ms. S. wants to combine the results into a summarizing index for analytical purposes, she might be tempted to assign each class a number and then calculate a mean for each respondent (Column b). However, performing arithmetic operations that yield mean values assumes that the data have been converted into interval data. To demonstrate the effects of this assumption, suppose the attitudes of P, Q, and R were measured by some other method that, indeed, did produce interval data (Column c). Note that the responses for the two types of measurement are the same according to their degree of specificity, where the class of 4s is equivalent to interval values of 3.5–4.5 and the class of 5s to values of 4.5–5.5. The means, which can be properly calculated from interval data, rank respondent R as highest and Q as lowest. In contrast, the ranking obtained from the converted ordinal data incorrectly ranked the respondents in a different order. This illustrates that it is quite possible that the assignment of numbers to essentially ordinal data without knowing the intervals between ranks can produce erroneous conclusions (also see "Ratings" below).

Table 3.1 An Illustration of Possible Effects of Data Conversion

Statement I. D.	a Responses by Respondents			b Assigned Numerical Values for Each			c True Interval Values		
	P	Q	R	P	Q	R	P	Q	R
I	SA	SA	SA	5	5	5	5.4	4.6	5.4
II	A	SA	A	4	5	4	4.4	4.6	4.3
III	A	SA	SA	4	5	5	4.0	4.6	5.2
IV	SA	SA	SA	5	5	5	5.4	4.6	5.1
Mean:				4.5	5.0	4.75	4.8	4.6	5.0

Ratio

Measurements produce interval data when the numbers assigned to phenomena are based on an arbitrary position for zero, so a comparison of two numbers as a ratio is not meaningful. For example, it is meaningless to declare that a place with 40°F has twice as much "temperature" as another with 20°F. In contrast, when zero is a meaningful assignment, the data are based on what is called a *ratio* scale or level of measurement. Distances measured from the Choros elementary school building illustrate this type since "a distance twice as far as another" conveys meaning. The differentiation between ratio and interval is not stressed here because both are expressed in numbers that can be statistically analyzed by techniques termed *parametric.* Without delving into all the properties of parametric statistics, it is sufficient to state that parametric data result from measurements at either the interval or the higher ratio level.

In summary, a fundamental question that must be considered before starting to collect data is this: What level of measurement is feasible and will it produce data that satisfy the analytical stage of the research? If the researcher plans to test a hypothesis using parametric statistics, gathering nominal data would be useless and a waste of energy. If, however, the research phenomenon can be measured only by nominal data, then an initial research plan calling for analysis by parametric tests will need to be altered to accommodate the lower level of available data.

Other Aspects of Measurement

In addition to deciding on the level of measurement, the researcher must consider some other aspects of the measurement task, namely, validity, reliability, precision, and accuracy.

Validity

Validity refers to the degree of coincidence between what is actually measured (which may be unknown) and what the researcher intends to measure. If the instrument (e.g., meter stick, questionnaire) does in fact measure what it purports to measure, it is said that it has a high degree of validity. For example, if Ms. S. states she is measuring the distance between two points, it should be easy for another person to verify whether she is or is not measuring that property called "distance". However, when she attempts to measure a concept called "attitudes toward riding school buses", other persons may doubt that her procedures of asking questions and tabulating responses and/or the specific instrumental set of questions are actually measuring what she thinks it is measuring. They may be right. Respondents may be reacting to her statements in a way that is not measuring attitudes toward school busing per se but is actually measuring something else, for instance, their feeling about taxes for school operations.

In many respects, the concept of validity introduces semantic conflicts. As an illustration of semantic or definitional controversies arising from measurements, consider the following rather absurd situation. Suppose a researcher decided to gauge people's "taste for pizza" by measuring the distance between their homes and the nearest restaurant selling freshly-baked pizza. The researcher could specify a procedure for measuring distances that would undoubtedly give good results, but the majority of pizza connoisseurs would disagree that the procedure actually provided a measurement of the designated phenomenon, that is, what they call a "taste for pizza." Often problems with the validity of measurements are of this nature (although not so obvious) because a researcher does not identify the phenomenon to be measured in clear and explicit terms generally accepted by others.

Some methodologists suggest seeking the opinions of experts to obtain their judgment about whether a specific instrument and procedure will measure the attribute for which it is intended. Another form of validity verification is by calculating an index of correlation between the results obtained from a new instrument and those from a proven one. However, these attempts to insure a high level of validity do not guarantee success because validity is basically a concept, a concept about the true quintessence of a phenomenon. It is impossible to verify completely the validity of a particular measuring procedure. In summary, the main goal for the amateur researcher should be a critical assessment of the question: "Am I really measuring what I want to measure?"

Reliability

Reliability is the extent to which independent researchers applying the same measuring procedures to identical phenomena obtain similar results. It refers to the probability that a procedure and instrument will provide a consistent measurement of the pertinent property, assuming the property itself does not change. A certain degree of reliability is always necessary for measurements because, if reliability is completely absent, no measurement can be accepted. The goal, of course, should be to develop procedures and instruments that generate data with high reliability.

Attempting to judge the reliability of a measurement is complicated when phenomena lack temporal stability. Independent researchers who measure phenomena at various times may obtain different results because (a) the measurement is unreliable, (b) the phenomenon changes, or (c) both unreliability and temporal instability exist. A few illustrations may help clarify how these factors occur in contrasting circumstances.

If Ms. S. asked one of her friends to go to Choros to "count vehicular traffic" along Highway #10, the friend's results might differ from those obtained by Ms. S. The difference reflects not only the variability in traffic flow at various times of day and night but also the vagueness of the measurement procedure as enunciated by Ms. S. She should have been much more explicit in describing the exact procedure for identifying and counting vehicular traffic. As is apparent, the task of specifying a measurement procedure is very similar to operationalizing a definition.

If Ms. S. uses a thin but long rope to measure distances, she might obtain different results for a particular earth distance because temporal variations in humidity and temperature would affect the elasticity of the rope. In this illustration, the phenomenon itself (the distance between places) does not change, and probably her directions for using the rope are fairly standardized. Therefore, it must be the variability of the instrument (the rope) which causes the inconsistency in measurements.

After giving the attitude questionnaire to selected Choros residents, Ms. S. thought about re-testing the same people to see if her questionnaire instrument was "reliable". Inconsistent results from a re-test would not have necessarily indicated unreliability, however, because the respondents would not have been the same as when they first took the test. They would differ because they had already seen the test once and they may have changed their ideas about buses. Thus, the factor of temporal instability would complicate her attempt to isolate the reliability factor. Instead, she might have tried to check on the reliability of the original questionnaire by including questions that nearly duplicate each other and thus provide a clue to their internal agreement and consistency (see Chapter 6).

Precision

Precision deals with fineness of the measurement scale and the exactness of the assignment of numerals. Ms. S. might report the distance between two places as 2½ blocks, 29 meters, or 28.74 meters. Each of these answers may be correct for the exactness used in measuring; the first one to the nearest half block, the second to the nearest integer meter, and the third to the nearest hundredth of a meter. They differ because they represent increasingly finer levels of precision. The concept of precision may apply to measuring any kind of phenomenon, including attitudes and other intangible attributes of human beings. It may also refer to the exactness in measuring locations, usually referred to as *locational precision*.

The degree of precision is partly related to whether data are continuous or discrete. When matching phenomena with a set of integer numbers, which produces discrete data, the level of precision is quite obvious. In contrast, when the assignment of numbers involves a continuous scale, the researcher must decide on an appropriate level of precision, for example, one that yields 28.74 as compared to 28.74366. The reader should remember that the terms discrete and continuous are also applied to the spatial characteristics of phenomena. Their usage should not be confused. An areally precise discrete feature, for instance, may be measured imprecisely as continuous data.

Accuracy

Accuracy refers to the degree to which the value assigned by the researcher agrees with the value that is "true" or without measurement error. The term is sometimes used informally to cover these other aspects of measurement, but the researcher should distinguish between accuracy and validity, reliability, and precision. Data may be accurate but unworthy in terms of the other three aspects. The measurement of distance to a restaurant serving pizza may be very accurate; but if the research problem concerns the "taste for pizza", the accuracy of measurement does not compensate for the lack of validity. Also, a series of measurements may be highly accurate, for example, in counting the number of "yes" responses on a yes-no questionnaire, but this does not guarantee that they are obtained from an instrument that produces consistent, and hence, reliable data. Furthermore, measurements may be accurate but imprecise (e.g., 2½ blocks). Conversely, measurements may be valid, reliable, and precise but yet be inaccurate. Consistent errors are usually associated with bias, an important topic that is discussed in Chapter 4 and in several subsequent sections.

Measurements of Visible Phenomena

A large share of geographic research deals with visible phenomena, and many of the traditional ways of gathering field data involve viewing features of the landscape. A common procedure is to go to the field, observe the spatial variations of selected phenomena, and map them as a record for subsequent analysis. Many phenomena are studied by this procedure, including such visible features as urban land use, rural settlement patterns, crop regions, and traffic flows. Because so many research problems are concerned with visible landscape features (NV and CV) and observable human activities (PV), it is essential that the procedure for measuring them be understood.

In one respect, observing visible features is simple because the data collector needs only to look—no expensive equipment or complex instruments are required. It would seem that almost anyone could go into the field with a list of categories that belong to a particular classification scheme (such as the various classes of land use), look at the features to be categorized, decide on the correct class for each individual, and record the information on a map or data sheet. Because of the apparent simplicity of the procedure, amateur researchers tend to overlook the potential frustrations involved with making reliable decisions. Conscientious researchers who strive to obtain reliable measurements will often have difficulty making decisions about the assignment of each individual to its proper class. It is frustrating to a painstaking, but ill-prepared, worker to realize part way through a project that two very similar individuals have been placed into different classes.

The next few paragraphs deal with ways of improving the measurement of visible characteristics of phenomena. They pertain to the task of measuring objects by assigning a class/numeral to each individual or, as commonly expressed, by assigning individuals to classes.

Identification

The initial step in measurement is to identify the individuals belonging to the set of objects that will be measured. This initial identification chore is a type of measurement in itself, so prospective fieldworkers must understand the nature of this fundamental task.

The Task

A basic measurement in data collection involves the *identification* of relevant individuals by dividing phenomena into two classes: those individuals that belong and those that do not belong to the set of phenomena to be studied. The set of phenomena to be observed and the procedures for measuring them should have been operationally defined. The task for the fieldworker is to decide which individual features and events fit those definitions. The procedural decision involves measuring at the nominal level because each individual is assigned a name. Basically, there are two classes with the names of "belonging" and "not belonging." In actual practice only the term for the defined phenomenon is mentioned (e.g., a vehicle) because the excluded individuals are no longer pertinent.

The task of field identification is illustrated by Ms. S. when she planned to count vehicular traffic on Highway #10. As noted previously (Chapter 2), she had to decide on an operational definition for the expression "vehicular traffic". This meant she had to determine which specific phenomena (tractors, cars, wagons, baby strollers, roller skates) fit her concept of "vehicles" and what events constitute "traffic". She decided that tractors (as well as several other vehicles, but not baby strollers and roller skates) did qualify for her definition of vehicles. Also, she stated that, to be measured, the vehicle had to move under its own power. She then had to decide which specific objects matched the definition of tractor, or other specified vehicles, and moved under their own motive power. That decision was the identification one.

When object X rolled past her, she had to make a decision about whether X did belong to the set of operationally defined individuals (vehicles) or did not belong. She would be able to make a meaningful decision if her previous experiences were adequate for recognizing X as a tractor (or other qualifying vehicle). The identification task for Ms. S. was not too difficult because most of her previous experiences with the objects constituting her operational definition were adequate to give consistent results. Likewise, she was confident that other researchers who used the same operational definition and whose background experiences included the recognition of phenomena called vehicles would identify the same data set.

Difficulties may arise if another person with an entirely different background attempted to use the definitions and directions prescribed by Ms. S. If Mr. W., who had never seen a tractor, attempted to use her definition of vehicles, he might be uncertain about whether a particular object moving along Highway #10 was the thing called "tractor" or not. A way of overcoming this handicap would be for Ms. S. to describe a tractor in terms which bridge her experiences and those of Mr. W. Such a detailed description can usually be provided for tractors and similar objects, but phenomena like "scenic beauty" and "architectural compatibility" often challenge the definitional skills of researchers. This illustrates that the measurement of visible phenomena, which seems to be accomplished by "just looking", really requires a carefully designed translation from an initial concept to an operational definition and a specification for measuring.

The task of identification requires the researcher to make subjective decisions about whether each specific individual being observed in the landscape does or does not belong to the set of objects, persons, events, or areas that were operationally defined as the data set. Often the decisions cannot be re-examined and verified at a later time, so they must be achieved correctly by the researcher at the time the data are collected. Some phenomena such as urban land use may not change quickly, so researchers can return to double-check field decisions they or others had made previously. In contrast the vehicular traffic on Highway #10 from 6 to 10 o'clock on April 29th can only be observed once. A mistake in measurement can never be detected or corrected later! All incorrect decisions will result in extraneous individuals and/or omissions. These may ultimately affect the analysis and research conclusions.

Techniques

Some procedures and techniques that may aid the researcher in making consciously reliable decisions can be illustrated best by simplifying the conditions affecting the decisions. Rather than considering vehicular traffic, therefore, attention should be focused on a task of sorting out red beads. Before an individual is a collection of small colored objects, which has successfully been separated into beads and non-beads. Now the task is to identify those beads that are red.

Maybe there is an instrument that quantitatively measures redness. All the beads with a color wavelength within the boundaries of red (say, between 610 and 700 nanometers) will constitute the desired assemblage of beads. That is a procedural solution, but the focus here is on making non-quantitative measurements like a researcher makes when viewing landscape features or events.

The task is to identify the beads that are red—just like a geographer might want to identify any visible feature such as a "farmstead" or "shoe factory". The difficult part of such a task is to deal with those beads having a color which is transitional to red. The transition from red to red-orange to orange, for instance, is gradual and no abrupt difference signals a boundary between red and red-orange. Without any guide for separating red beads from non-red beads a different set of "red beads" may be selected by different persons and even by an individual at different times. These differences reveal an undesirable condition in data collecting, namely, the lack of reliability.

What is needed is a technique for identifying red beads that will provide a fairly precise visual measurement of the boundaries of redness. One technique is to select a bead that represents the ideal, the norm, or model of redness. Then all other beads can be compared to the model bead and those that deviate too much from ideal redness will be excluded. In this manner, a boundary to the class of beads that are red will be established by the amount of deviation from the norm. The advantage of this procedure is its manifestation of an abstract concept (redness) by a concrete object (the model red bead). A limitation to this technique is that it depends on estimating the amount of deviation from the ideal. Stated otherwise, a researcher must judge how far a particular individual bead is along the transition from ideal redness to non-redness.

A second technique for measuring membership to the bead assemblage requires the selection of two beads that are on the boundaries between red beads and non-red beads. Because the attribute of redness is one dimensional, other beads can be classified as red or not by observing on which side of the two boundaries they belong. In this case it is necessary only to consider the question: "Is the observed bead more red or less red than each of the standard boundary beads?" This is an easier comparison than one necessitating a quantitative judgment of distance, i.e., the deviation from the norm. However, when observing phenomena with several relevant attributes (see below), it may be a little more difficult to compare the multiple boundaries than to make comparisons with a single model individual/element.

The first task of identification was simple because it measures only the attribute of color. Suppose the beads vary in size as well as color, and the objective is to select beads that are red and marble-size. Do not worry here about the definition of marble-size per se but rather concentrate on the procedure for achieving consistency in applying whatever definition is accepted. Note that with two characteristics to observe, it may be necessary to select two standards for the model and/or several for the boundaries if no bead combines color and size in exactly the desired combination. If a separate bead is selected for each attribute for both norm and boundaries, the identification process could involve as many as six comparisons. Six would result if each bead to be classified were compared with the bead having an ideal redness, the one having an ideal marble-size, the two beads representing the boundaries of redness, and the two beads typifying the upper and lower limits of marble-size.

If a third characteristic such as shape is considered pertinent, the comparisons become even more complex. Most phenomena being measured during actual research projects involve relevant characteristics that number even more than three, so the goal of standardizing identity decisions is a difficult one to reach. For example, Ms. S. classified dwelling units in Choros. In most cases she was able to identify the buildings and portions of structures that qualify as dwelling units. Nevertheless, some structures remained difficult to classify quickly, and several questions bothered her while in the field. Does the upstairs portion of the house where elderly Mrs. V. lives qualify as a dwelling unit separate from the rest of the house where her son and his family live? Is that dilapidated unoccupied house still a dwelling unit? Is each motel room a member of the set of dwelling units in Choros? These questions suggest several factors that are related to the operational definition and field identification of dwelling units. Each factor should have been considered in an objective and prescribed manner to achieve a reliable measurement.

Classification

Classification is the process of objectively assigning a nominal designation to each individual on the basis of specific criteria. It is a basic aspect of most scientific research in all disciplines because it involves the measurement of phenomena, an organization of the data, and a preliminary testing of relationships.[2] Classification is very important in geography, not only because of the traits geography shares with other disciplines that use scientific methods, but also, because many phenomena are observed only visually. As a result, classification becomes the method for measuring the phenomena. Classification is especially important in the myriad of land use studies where parcels of land are looked at and then assigned a class designation.

Because identification places individuals in one of two classes, namely, a belonging class and a not-belonging class, it can be considered a type of classification. The term identification is restricted in this text, however, to refer to the initial classification process that separates the members of the study set from all other individuals that are observed in the field. The subsequent divison of the study set into differentiated groups is termed classification. The role of each of these terms is clarified more by a diagram (fig. 3.1), which shows the process of moving from the general conceptual term of a phenomenon to the recorded data. First the researcher translates the general term into an operational definition that aids in making a consistent judgment in the field. Then while observing the mass of individuals in the field and following a measurement procedure or defined operation, the researcher separates the individuals that match the definition from those that do not. This field operation is the identification step. Measuring these selected individuals, when accomplished by looking at them, invariably involves classification. This task requires the researcher to convert an organizational concept into specific criteria for matching individuals with classes and hence producing a set of data.

Approach

The goal of classification is the grouping of similar individuals together to form a few general classes. This may be accomplished by either of two approaches: by the aggregating approach or by the dividing approach. The *aggregating* approach begins with all the individuals that are to be classified (let the number of them be n). Those individuals that are most similar according to a specified criterion, or several discriminating criteria, are joined together to create a group or class of individuals. If fewer classes are then desired, the small groups may be joined together to create larger and more generalized classes. The procedure for aggregating classes by joining similar individuals into groups (classes) is normally accomplished during the analytical stage of research by statistical techniques which can cluster quantitative data. Consequently, this aggregating approach is not directly related to the task of designing a classification scheme or classifying visual phenomena in the field—most of the remaining comments in this section pertain to the second approach.[3]

The *dividing* (divisional) approach commences conceptually with all the individuals being regarded as members of a single group, which is then divided into a number of classes. The classes must be defined in such a way that each individual belongs to one, and only one, class. The assignment of each individual to its proper class is achieved on the basis of one or more criteria, but always subject to the provision that the criteria define mutually exclusive classes.

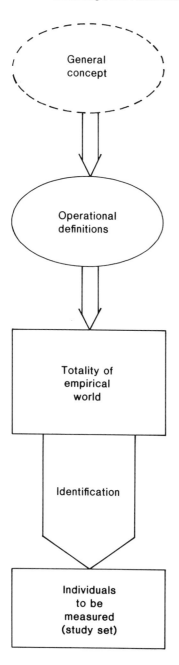
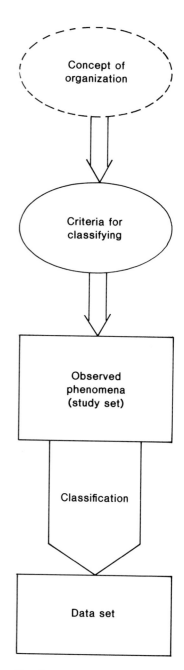

Figure 3.1. Processes of identification and classification

If more classes are desired, usually for the purpose of reducing the diversity among individuals assigned to each category, the classes can be subdivided. If these subclasses are subdivided again, another level of classification is achieved and the scheme possesses a hierarchy of classificatory levels. In this respect the two approaches are opposites. The divisional approach begins with one general class composed of all n individuals, progresses toward more specificity as the level of classification changes, and theoretically terminates with n one-member classes. The aggregating approach, on the other hand, begins with n unique classes, progresses toward more generality as the hierarchical level changes, and theoretically terminates with a single n-member class.

The foregoing description of divisional classification included two expressions that need further clarification. First, the divisional approach commences "conceptually with all the individuals", but in actual practice the classifier seldom observes all the individuals to be classified at the time the scheme is developed. This means the researcher can only hope that the scheme which is prepared prior to field classification will accommodate all individuals. Success in anticipating the many variations in the sample population can be improved by watching for such diversity during a field reconnaissance. The second clarification concerns the progression of a hierarchical scheme to the level where it "theoretically terminates" with complete uniqueness. This extreme level of classification does not make sense because complete individuality is equivalent to no classificatory measurement. In actual practice, the appropriate hierarchical level must be decided by the researcher who should consider several factors incorporated into any classification scheme.

Choice of Scheme

A researcher who plans to measure phenomena by using a classification scheme must decide whether it is better to accept a scheme already established by previous scholars or to create an original one. A reason for using an established scheme is that it may have been applied successfully in a variety of situations, giving the researcher confidence in its utility. Furthermore, since it has been used elsewhere, the results should all be comparable, a fact which enhances the value of any new research.

The merits of using an established classification scheme can be exemplified by the land use classification system utilized by several U.S. governmental agenices.[4] This system was developed after numerous experts examined various land use and economic classifications (e.g., the familiar Standard Industrial Classification scheme); thus the authors of the land use system combined considerable experience into its construction. It is a hierarchical system with 9 major categories, 67 classes at the second level, 224 at the next subdivision, and 772 classes at the most specific level. For example, one of the major categories is called "resource production and extraction"; one of its subdivisions is "agriculture", which includes as one of its subdivisions "farms and ranches"; this latter class in turn includes among its subdivisions the class of "sheep farms and ranches". These categories are standardized, making it possible for different field workers in diverse regions at various times to classify similar land areas into the same class.

There are also reasons why researchers may choose to design original classification schemes rather than using ones already available. A primary reason is the possible lack of any existing scheme that measures phenomena in the manner that researchers require to solve their research problems. Although novice researchers should investigate existing schemes that may pertain to the relevant phenomena in their studies, they should not be surprised if they do not find one that is suitable.

Another reason for seriously considering the development of an original scheme is the heuristic value associated with constructing a suitable classificatory system. As emphasized in this chapter, making classificatory decisions is a type of measuring that demands much effort to maintain reliability. Just carrying an established scheme such as the Standard Land Use Classification Coding Manual into the field does not guarantee consistency in classifying the phenomena. A thorough knowledge of a scheme's structure is also needed for assuring reliable measurements. The researcher who creates an original scheme, whether it is ultimately used or abandoned in favor of an established one, usually obtains more reliable results in the field than a person who has not pondered the various decisions that go into making a classification scheme.

Criteria

The criteria for defining classes and assigning individuals greatly influence the utility of a classification scheme. In effect, specifying the criteria for defining classes is analogous to stating operational definitions because it designates the attributes of the phenomena that are most significant. In general, the decision about which criteria are relevant should be based on the research problem and the hypothesized relationships among variables. Normally, an experienced researcher is able to make a logical choice about the characteristics that will best discriminate among individuals and, therefore, serve as classificatory criteria.[5]

Closely related to the question about which characteristics should be used in a classification scheme is the question of number. The researcher must decide how many criteria are necessary to differentiate individuals effectively. Whenever a researcher is uncertain about the characteristics that will be most indicative of the hypothesized relationships, it might seem like a good idea to include numerous criteria to insure that no important ones are omitted. An increase in the number of criteria, however, produces more complexity in the scheme and possibly more difficulties in the field when assigning individuals to their proper class.

The conflict between the advantages of a few criteria in a simple scheme versus those of a scheme that is thorough but has many criteria is partly resolved by constructing a hierarchical classification. By having several classificatory levels, some of the advantages of both simplicity and thoroughness may be achieved. Although a hierarchical scheme may not be simple in its entirety, the classification task at each level may remain fairly easy.

Some advantages of the hierarchical scheme can be illustrated by returning to the controlled conditions of the beads. Assume that this time the problem is to separate the objects identified as beads into classes. The decision has been made that color, size, and shape are meaningful criteria (and that composition, age, ownership, and other possible attributes are not relevant). Suppose that the colors can be divided into four classes, size into three, and shape into three. This would produce thirty-six possible classes ($4 \times 3 \times 3$) based on the various combinations of the three criteria. Comparing each individual with all the thirty-six possible categories to see which one matches best would be an extremely tedious field task. If, in an attempt to simplify the task, only color is selected as a discriminating criterion, the complexity of the problem is reduced. To classify the beads now requires only four comparisons. This simpler scheme, however, does not utilize available information about size and shape. Using a hierarchical scheme retains the advantages of having three criteria while also reducing the number of comparisons required at any one level. This is seen by looking at a dendrogram, which shows all the classes in a tree-like diagram (fig. 3.2). At the first division a decision must be made as to which of the four color classes a particular

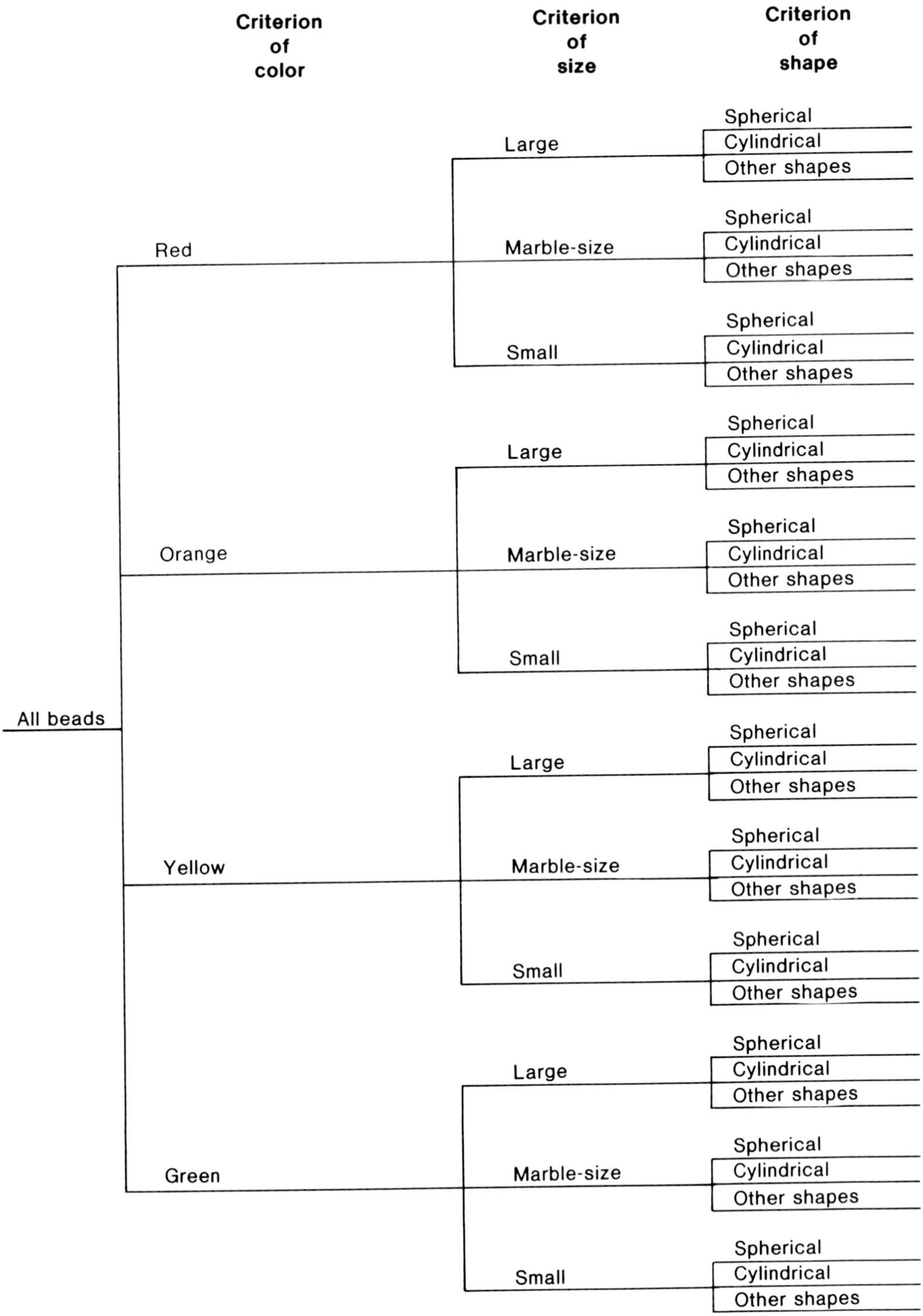

Figure 3.2. Dendrogram of a scheme for classifying beads

bead belongs. Assume it is orange. At the next level this orange bead is compared with the three size classes. After classifying its size (say, small), three possible shape categories are next examined. In this manner, the bead is measured, nominally in terms of its color, size, and shape without having to make more than ten (4 + 3 + 3) comparisons.

One of the subtle disadvantages of a hierarchical scheme is its implied ranking of the criteria, even though presumably they should have equal discriminatory power. As illustrated by the classification of beads, it would appear that more importance is given to color than to size and shape. For example, the scheme suggests that an orange-small-spherical bead and an orange-small-cylindrical bead are more similar (i.e., less widely separated) than are the two beads classed as red-large-cylindrical and green-large-cylindrical.

Classes

Classification schemes are also affected by the number of classes and the relationships among those classes. The decision on the number of classes is important because it directly affects the precision of measurement achieved by the classification scheme and the difficulty of the classification task in the field. If the researcher chooses only a few classes per criterion, then the variation within classes tends to be large. As a result, the scheme is simple (for a given number of criteria) but precision is usually sacrificed. If a scheme has many classes, then each class will be quite precise because internal variation is restricted. The "costs" for achieving this increased precision through more classes are the greater complexity of the scheme and the additional number of borderline judgments to be made. The greater complexity is not necessarily a disadvantage, but it may be more difficult to explain the scheme to other persons. The added number of boundaries between adjoining classes, however, will require more time in the field to make classificatory decisions. Thus, the question about the number of classes in a classification/measurement scheme presents the researcher with a dilemma of conflicting objectives. To increase the precision of measurement, the scheme should contain more classes. To reduce the number of boundaries and hence the number of classificatory decisions, the scheme should possess fewer classes. The way a researcher resolves this dilemma is one of the fundamental issues of measurements.

The number of boundaries between classes depends partly on their relationships. If the classes per criterion can be ranked, then the number of possible boundaries is greatly reduced. For example, if the four color classes can be meaningfully ranked (say, in the usual sequence: red, orange, yellow, green), then the researcher needs to worry about three boundaries, namely, the red-orange, the orange-yellow, and the yellow-green ones. Likewise, if five classes of some phenomenon X can be ranked, comparisons need to be made only for four boundaries. Not only does this make the measurement task easier, but it provides ordinal data. If, however, there is no logical ordering of the classes, then the boundaries between all possible pairs need to be checked. Unless a plausible ranking for the three shapes of beads exists, each bead must be compared with three boundaries (cylinders-spheres, cylinders-others, spheres-others). Similarly, to classify individuals of some phenomenon Y into five categories that cannot be ranked will require ten comparisons.

These comments about the number of classes and their relationships can be expanded to situations with multiple criteria. Furthermore, rather than sorting beads in isolation, researchers normally must cope with the complexities of phenomena in their environmental contexts. When all the options in designing a classification scheme are applied to the real world, the magnitude of decision-making becomes almost overwhelming. As an example, consider some of the questions that might occur while constructing a scheme to classify rural land use. What is meant by rural land use? How can "rural" and "land use" be operationally defined to communicate a consistent meaning of the concept? Does it include greenbelts around metropolitan areas, military lands in isolated areas, and farms within a city's limits? Will the features that define rural land use always be recognized when seen in the field? What technique can be used to identify phenomena that seem to be transitional between rural-land use and non-rural land use? What categories of land use will best differentiate rural areas? On what criteria will land use classes be based? Will the criteria be primarily current economic utilization? How should areas that visually appear to be waste land be classified? Should ownership, especially a differentiation between public and private, be a criterion? What other criteria might be important? If several criteria are involved, should they be ranked and incorporated into a hierarchical scheme? How many classes are needed? Should they be general ones such as agriculture, residences, transportation, recreation, mining, manufacturing, water areas, and vacant lands? Should finer divisions be included? For example, should the agriculture class be divided by production types such as grains, fibers, fruits, vegetables, dairy, livestock, poultry, and miscellaneous others? Is there any characteristic by which these classes should be ranked to produce ordinal data?

These certainly are not all the questions that might be asked when establishing a classification scheme for measuring rural land use; but they typify the kinds of questions and decisions a researcher faces before going into the field to collect data. Certainly they reveal that the task of gathering geographic facts involves more than "only looking" at the phenomena. The collection of data about visible phenomena requires conscious consideration about how the phenomena will be measured. Only after making these decisions and establishing a classification scheme is the researcher ready to go to the field to collect the data. Comments about procedures and techniques that can be applied while actually making classificatory measurements in the field are in Chapter 5.

Measurements of Non-Visible Phenomena

Field data may be observed through any of the human senses, but visual observations and listening to human utterances are the primary media. This section deals with the measurement of phenomena having the form of the verbal responses, both spoken and written, provided by individual persons. The emphasis continues on ways these forms can measure the properties of individuals.

In some respects it would seem that collecting valid data from respondents who already "know the answer" should be easier than relying on the researcher's ability to measure phenomena by only viewing them externally. Success does not come easily, however, because the collection of information depends on (1) the respondents' abilities to measure requested knowledge and/or attributes about themselves and (2) the accurate communication of facts between the researcher and each respondent. This second aspect, the accurate communication of information, is the topic of a subsequent chapter (Chapter 6); here the focus is on the measurement aspect. In this section it is assumed each respondent possesses the requested information (e.g., about age, marital status, choice of location for a new building, feelings about children walking rather than riding to school) and is willing to share the information with the researcher. By what techniques can the researcher use these responses to measure various properties, especially those involving attitudes, beliefs, and feelings?

Nominal Measurement of Non-Visible Phenomena

Nominal data are produced when the respondent is asked to indicate which of a choice of classes or potential answers is the correct one. There may be just the two classes of a dichotomous choice (see Table 3.2, #1 and #9) or several classes presented in a multiple-choice format (Table 3.2, #2). A check list that allows more than one answer is essentially a series of dichotomous choices with a yes-no response option for each of the items listed in the set (Table 3.2, #5). A modification of the check-list format is a scattering of potential answers to which the respondent must reply (Table 3.2, #7). Nominal data can be obtained also from open-ended questions if the responses are analyzed for their content (Chapter 8).

Table 3.2 Various Kinds of Questions

. . .

1. Are you a resident of Choros?
2. What is your present marital status?

Yes _____ No _____
Single and never been married _____
Married _____
Single but previously married _____

. . .

5. By what different methods will your child go to school next year? (Check all that apply.)

_____ Walking
_____ Riding a bicycle
_____ Riding in our family car
_____ Riding in another family's car
_____ Other (list) _____

. . .

7. Encircle the words that you feel best describe the existing elementary school building in Choros:

ADEQUATE   OLD
WELL SITUATED   DETERIORATING
STURDY   UNATTRACTIVE
                                              CHARMING
POORLY LOCATED
UNSATISFACTORY

. . .

9. Do you believe Choros should build two new elementary school buildings?

Yes _____ No _____

. . .

Below are fifteen statements. Respond to each statement by encircling the answer that best describes your feeling about the statement.

11. The Choros school district should replace the old elementary school building by constructing new facilities.

    Strongly Agree   Agree   Undecided   Disagree   Strongly Disagree

12. The present elementary school building is in a poor location.

    Strongly Agree   Agree   Undecided   Disagree   Strongly Disagree

Table 3.2—*Continued*

13. Elementary children should be expected to walk to school provided the distance does not exceed one mile.

    Strongly Agree   Agree   Undecided   Disagree   Strongly Disagree

14. The existing elementary school building would be satisfactory if it were renovated.

    Strongly Agree   Agree   Undecided   Disagree   Strongly Disagree

. . .

Below are twenty pairs of opposing words; each pair is separated by five spaces. On each line check the space that best describes your feelings about educational facilities in Choros.

31. good _____ _____ _____ _____ _____ poor
32. old _____ _____ _____ _____ _____ new
33. pleasant _____ _____ _____ _____ _____ unpleasant
34. well supported _____ _____ _____ _____ _____ unsupported
35. worthless _____ _____ _____ _____ _____ valuable

. . .

Below are twenty-two statements. Indicate your agreement or disagreement with each by marking an X on the line below the statement.

61. The Choros school district should provide bus service for all elementary students.

    Strongly   Strongly
    Agree _____ Disagree

    . . .

Most respondents understand a request for indicating answers in the form of two or more mutually exclusive choices. It is the confusion in wording questions and the options that tends to produce distorted or unduly filtered data. Suggestions for improved wording are given in Chapter 6, but a few questions selected from a schedule used in Choros (Table 3.2) may expose the kind of measurement weaknesses that can occur. For example, the choices provided for marital status (#2) display an aspect of imprecision because the "single but previous married" option can include both "single and divorced" and "single and widowed." This imprecision may not be a serious fault in some research problems, but the designer of the schedule/questionnaire needs to consider the potential diversity that can occur within each response choice (class).

Reliability of answers from the check list (#5) is not jeopardized so much by the options as it is by the requested information. Because the respondent must ponder a hypothetical (future) course of action, there is greater probability of obtaining unreliable data than if the information requested is on past behavior. In other words, the kind of knowledge that is being requested, namely, a prediction on future action, is normally answered in a less consistent manner than facts about previous actions.

The assumptions of the researcher and those of the respondent may differ and thus produce data that lack much validity. The questioner may intend to ask Choros citizens (#9) about only two building proposals: erecting two medium-sized school buildings (yes) vs. constructing one large building (no). However, numerous respondents in their own minds may compare "two new buildings" (yes) with "not two new buildings" (no), a category including both "one new building" and "no new building." The question, then, measures something other than what the questioner intended and thus possesses low validity.

Ordinal Measurements of Non-Visible Phenomena

A dichotomous choice (e.g., Table 3.2, #1 and #9) does not provide a precise measurement, unless the property is already polarized. If the characteristic occurs as one of two obviously distinct classes (resident/non-resident; male/female), a dichotomous choice provides precise data. If, however, the characteristic varies along a continuum (length of residency; feelings about two school buildings), then another level of measurement is desirable. Opportunities for achieving more precision are obtained by asking respondents to classify their answer according to options that are ranked; that is, each respondent chooses the proper category among a set of ordered choices. Three forms for eliciting ordinal data are the Likert-type statements with associated response classes, the semantic differential with its accompanying ranked categories, and card-sorting that requires ordering elements into a set of classes.

A *Likert scale* consists of a set of statements to which the respondent reacts by marking one of several ordered classes.[6] The entire set of responses often is designed to provide a quantitative measurement (see below), but statements considered singly produce data ranked along a scale varying from a favorable pole (class) to an unfavorable pole. A five-class scale might be verbalized by the terms "strongly agree," "agree," "undecided or indifferent," "disagree," and "strongly disagree" (Table 3.2, #11–14). The number of classes is normally more than three because three classes are not much more precise than a dichotomous choice with a class called "undecided/I don't know" in between. On the other hand, when the number of classes exceeds seven, respondents may have difficulty discriminating among adjacent classes, and reliability is diminished. Five is the most common number of classes. An odd number of classes allows for a neutral response; an even number forces a decision toward one end of the scale. There is danger of inaccurate measurement if respondents who have no opinions are forced to indicate a preference toward one pole. This suggests an advantage of an odd number of classes. However, some issues demand making a decision and the inclusion of a neutral category only allows indecisive persons to avoid committing themselves.

Another bipolar scale on which respondents indicate attitudes by marking a category is the *semantic differential.*[7] The form consists of pairs of adjectives which have opposite meanings but both of which pertain to a general concept (Table 3.2, #31–35). If more than one concept is measured, then each has its own set of bipolar pairs. Since the technique is used to measure an abstraction, validity is especially important. It is often difficult, however, to know if a respondent is reacting to the concept that the researcher has in mind when using only descriptive words.

Card sorting requires respondents to arrange elements, which are usually cards with a word or statement written on each, into a set of ranked classes. In a structured version of this technique, the respondent is asked to classify each word/statement (card) by placing it in one of several (e.g., five) classes previously designed by the researcher. In an unstructured version, the respondent must decide the number of classes that seem to best measure the variation in the elements, as well as arrange each element (card) into its proper class.

Parametric Measurements of Non-Visible Phenomena

Many open-ended factual questions produce numerical data directly. What is your age? How long have you lived here? How many children are in this family? How often do you drive your children to school in the family car? By asking the respondent a quantitative question about a known fact the researcher can obtain interval and ratio data.

It is difficult to measure attitudes, beliefs, and feelings with a direct question. It would be rare, indeed, for a researcher to obtain a quantitative response to the question: "How religious are you?" Techniques for generating numerical data can be considered under the titles of ratings, indexes, and scales.

Ratings

One technique for producing parametric measurements attempts to coax numerical values from respondents by having them match each attitudinal response with a number and thus produce a *rating.* They are asked to evaluate or "measure" specified properties along a continuum from one extreme to its opposite extreme or pole. Usually the two extreme positions are described to guide the respondent, but such explanations may be omitted if it is desirable to have respondents envision their own extremes. A popular request takes this form: "On a scale of 1 to 10, how do you rate. . . ?" This kind of measurement was also illustrated when Dr. A. asked a group of teachers to rate various facilities in the school building. The teachers were requested to assign a numerical value from 0 to 20 for each of several conditions, for example, the condition of the plumbing, the quality of lighting, and the adequacy of the classroom space. Thus, each teacher had to match personal perceptions of each characteristic with an unspecified or open-ended set of numbers between 0 and 20.

Ratings often suffer because of uncertain validity, low reliability, and varied precision. If ratings are used for poorly defined phenomena, the inadequate communication between the researcher and respondent may jeopardize validity. For example, it is difficult to know whether the attribute called "adequacy of classroom space" is interpreted the same by Dr. A. and each teacher. Does a rating of 13.5 by two teachers for the same place at the same time indicate equivalent evaluations on the same characteristic?

Also with ratings it is difficult to achieve a high level of both precision and reliability. Teacher J scored the facilities only as multiples of 5 (i.e., essentially as classes with 0, 5, 10, 15, and 20 being class midpoints). By using such broad ranges for each score, teacher J was able to be fairly consistent in assigning scores to various rooms; but the ratings were rather imprecise. In contrast, teacher K rated the facilities to the nearest tenth of a unit (e.g., 13.5, 7.2). Although teacher K used a more precise scale than did teacher J, probably the total set of ratings for various rooms is internally less consistent than the scores assigned by J. In any case, the fact that teachers J and K used different levels of precision complicates the task of interpreting the data.

A modification of open-ended ratings interprets the responses to Likert-type statements as if they were interval data. Stated otherwise, the researcher provides the respondent an opportunity for rating a set of statements by using only discrete values at equal intervals. Thus, numbers, as well as descriptions, are assigned to classes. Typically the possible responses are these: 1 = strongly agree, 2 = agree, 3 = undecided, 4 = disagree, and 5 = strongly disagree. As noted above in the discussion about ordinal and interval levels of measurement, it would seem that parametric statistics are inappropriate for analyzing data generated from ratings based on numbered "classes." Some statisticians however, claim that the numbers can be interpreted as interval data in the same manner as other ratings that are not restricted to integers.[8]

Another way of producing interval data from Likert-type statements is by having the respondent mark a linear scale (Table 3.2, #61). By making the line a "standard" length (e.g., 100 mm), the researcher can measure the distance between the respondent's mark and either pole and thus calculate the mark's score. A primary disadvantage to this form of measurement is the implied precision. An X on a line can be interpreted as a precise point (unless it is scrawled across a large portion of the line), but it is doubtful that most persons have the ability to measure their attitudes that precisely. If the precision that is implied by the mark on a continuous line is accepted, then inaccuracy and unreliability in the measurement become highly probable.

Indexes

An *index* is a cumulative score obtained by combining the responses for separate components of the phenomenon being measured. Combining the separate scores to form an overall index introduces a procedural aspect that requires a subjective decision about whether the scores should be added, first weighted and then added, or mathematically manipulated in any other way to correctly represent the property being measured. In rare cases one may have a theoretical basis that will justify combining separate ratings in a manner other than by equal-weight addition. Usually indexes are summations of several separate ratings with each having equal arithmetic importance.

Even if statements are justifiably combined, there is no guarantee that each statement contributes equally to an overall concept. For example, of the four statements dealing with the Choros school building (Table 3.2, #11–14), it may be that most respondents have attitudes about children walking to school (#13) that differ from their attitudes about the existing school building (#11, #12, and #14). If, indeed, there are two different concepts (or, components to a general concept) being held by respondents and being reflected in their answers, then it would be misleading for a researcher to combine the responses into one index which purports to measure a single concept. On the other hand, maybe all four (#11–14) do manifest aspects of only one concept (component) held by most respondents. How can a researcher learn which of these two is the true situation? One way is to analyze the results of a pre-test.

A pre-test is similar to a "pilot project for a questionnaire"—it is a trial run for the purpose of determining whether a questionnaire will effectively measure what the researcher hopes it will do. There are many difficulties in designing a valid questionnaire and the pre-test provides an opportunity to check on many aspects (as is discussed more in Chapter 6). Included in the benefits of a pre-test is its potential utility in constructing an index. A pre-test provides a chance to obtain responses from a preliminary sample of persons, analyze their replies, and consequently decide whether the original statements seem to measure what they are supposed to.

An analysis of replies to Likert statements consists of tabulating the response pattern among various statements and the composite index to isolate duplications, inconsistencies, and irrelevant measurements. Specifics can be demonstrated by looking at the responses Ms. S. obtained on a pre-test (Table 3.3.a). She had twelve persons, A-L, respond to a set of nine statements, I–IX, by rating each statement on a five-class scale varying from "Strongly Agree" to "Strongly Disagree." She then quantified their responses by assigning a 1 to extremely negative replies and a 5 to the positive end of the scale. This step can be clarified by examining statements #11, #12, and #14 (Table 3.2). Ms. S. regarded the positive form of one attitude as "wanting a new school building in a new location." Therefore, she scored statements #11 and #12 as positive ones, even though #12 is worded in a negative manner. In contrast, for statement #14 she rated a response of "Strongly Disagree" as a 5 because the positive pole indicates a negative attitude toward a new location.

After tabulating the responses, she added the nine ratings for each respondent to obtain an overall index (Table 3.3.a). Then she ranked the respondents according to their indexes from highest to lowest and re-tabled the replies (Table 3.3.b). The summed results indicate that the respondents vary. The sums of their scores range from 38 for respondent C to 19 for H. Because of this variance, she can conclude that her statements were probably effective in measuring

Table 3.3 Analysis of Responses for Constructing an Index Score

a. Responses on a Likert-Type Scale (Numbers Obtained from a Rating 1–5)

Statement	A	B	C	D	E	F	G	H	I	J	K	L
					Respondents' Scores							
I	4	4	5	2	3	4	4	1	5	3	5	3
II	4	3	5	1	1	3	3	1	4	2	4	2
III	5	4	2	4	3	1	4	5	5	4	1	2
IV	2	1	3	1	1	1	1	1	2	1	2	1
V	4	4	4	4	4	4	4	4	4	4	4	4
VI	4	3	5	3	3	3	3	3	4	3	4	3
VII	3	2	4	1	1	2	2	1	3	1	3	1
VIII	4	4	5	2	3	4	4	1	5	3	5	3
IX	5	5	5	3	3	5	4	2	5	4	5	4
Sum	35	30	38	21	22	27	29	19	37	25	33	23

b. Re-Organized Responses

Respondent	I	II	III	IV	V	VI	VII	VIII	IX	Tentative Index (Sum)	Final Index
				Scores by Statements							
C	5	5	2	3	4	5	4	5	5	38	22
I	5	4	5	2	4	4	3	5	5	37	19
A	4	4	5	2	4	4	3	4	5	35	18
K	5	4	1	2	4	4	3	5	5	33	19
B	4	3	4	1	4	3	2	4	5	30	15
G	4	3	4	1	4	3	2	4	4	29	14
F	4	3	1	1	4	3	2	4	5	27	15
J	3	2	4	1	4	3	1	3	4	25	11
L	3	2	2	1	4	3	1	3	4	23	11
E	3	1	3	1	4	3	1	3	3	22	9
D	2	1	4	1	4	3	1	2	3	21	8
H	1	1	5	1	4	3	1	1	2	19	6

respondents' differences according to one or more characteristics. To build an index based on a single concept, however, she wanted to be reasonably certain that all nine statements pertain to that concept. By what procedure did she obtain some clues about the cohesion of these statements around a single concept?

The basic technique Ms. S. used was to examine the variations of scores in each column of the re-organized table (Table 3.3.b). She noted that Statement I evidently measured "something" that varies among the twelve respondents because their responses range from 5 to 1. Because the scores for Statement I covary with the Tentative Index, she believed that the "something" is a component of the overall concept. Likewise, Statement II seems to measure variation in some component that is related to the one assessed by Statement I but not identical to it. Statement III also measures variation in something, but whatever it is must be irrelevant to the phenomenon being measured by most of the other statements because it is poorly correlated with the sum of scores (Tentative Index). Respondents who measured high on the tentative index, such as C and I, differed on Statement III; and respondents H and D, who ranked low on the overall index, rated Statement III high. She decided, therefore, to discard Statement III from her revised version.

Statement IV does not show a large variation among the respondents because several of them are 1. However, within any set of statements it is valuable to include a few statements that measure polar positions, so she retained Statement IV. In contrast, the responses to Statement V show no variance. She discarded this statement from her final version because it contributed nothing to measuring variation in individuals. Although the limited variation in Statement VI resembles that in Statement IV but at the other polar position, it has less utility because of the predominance of 3s, which is the score indicating uncertainty or lack of knowledge. With so many respondents unable to provide an opinion on this statement, it has limited value. Ms. S. decided to exclude it from the version she plans to utilize for collecting data.

In the remaining three columns the response pattern for Statement VIII raises questions because it duplicates the replies for Statement I. If she wants to give double weight to this component in the index, she will retain both Statements I and VIII. If she wants to use Statement VIII as an indicator of respondents' consistency by comparing it with Statement I (in which case she will not include both scores in the index sum), she will not discard it. Otherwise, she will exclude Statement VIII along with Statements III, V, and VI.

The item analysis provided Ms. S. with a basis for judiciously deciding which statements do and do not contribute to a single concept and thus may be combined into an index. In her case the final index scores, which range from 22 to 6, were summed from the responses to Statements I, II, IV, VII, and IX. These statements are the ones she will use to collect the data that can be combined into a quantitative index to measure variations among individuals.

Persons who construct an index in this manner should bear in mind its limitations. One is its lack of any indicators of inadequacy in measuring a specific concept. The technique reveals whether certain statements hang together as they reflect some concept, but it does not deal with the issue of validity nor with additional statements that may be necessary to expose all aspects of the phenomenon being measured. Another limitation in using this kind of index is its lack of information about intensity of feelings. For example, an index of 15 for both respondents B and F does not reveal that B is very opinionated and feels strongly about each answer while F is rather indifferent to the issues expressed in the set of statements.

Scales

Another technique that provides interval data requires the construction of an attitude *scale*. This procedure attempts to rank statements in such an order that the response pattern reveals the respondent's position along a continuum. The statements themselves may only generate nominal data such as yes-no responses, but the ordered organization of the statements provides a measurement that is quantitative.[9]

The ranking of statements is the difficult step, so an illustration may help in understanding the procedure. Consider the following three statements:

    I. School buses should be run for those students who live more than one mile from the school and whose parents report a transportation hardship.

    II. School buses should be run for all students who live more than one mile from school.

    III. School buses should be run for all students in Choros.

If respondent K answered Statement I with "no," then there is no reason to inquire about the additional statements because their responses would necessarily be negative. If respondent L said "yes" to Statement I but "no" to Statement II, this indicates a different position along a measurement scale from respondent K. Likewise, a response pattern of "yes-yes-no" by respondent M for Statements I, II, and III respectively indicates a difference from that of respondents K and L. If respondent N replies affirmatively to all statements, this is yet another position along the scale that measures attitudes about running school buses. Note that a "yes" for Statement III implies the responses for the other statements *if* they are properly scaled along a single conceptual continuum (or dimension). Therefore, a score can be attached to each statement as a measurement of intensity or position along the continuum.

All of this depends, of course, on correctly ranking the statements according to the property being measured. The task of ranking statements may be accomplished by two different approaches. One depends on a panel of knowledgeable judges to make the decision. First, a large collection of statements (i.e., much larger than the set of three used in the illustration above) that pertain to the phenomenon are given to each judge for categorizing into several ranked classes. Commonly the number is not fewer than seven or more than eleven. The judges will undoubtedly differ somewhat, but the researcher hopes they will generally agree on the ranking of several statements. Those statements that are evaluated similarly by all judges are selected from each class by the researcher to form the final set. The researcher records the ranked position of each selected statement because this information forms the measurement scale. However, the order of statements is randomized on the final questionnaire so respondents will not detect the ranking envisioned by the researcher and judges.

The second approach does not depend on a group of experts to judge the proper order of statement. It uses the results of pre-test to determine how sampled respondents evaluate the statement. To obtain a ranking of statements, the researcher takes the results from a preliminary run and analyzes them by a procedure sometimes called scale analysis or the scalogram method. This analytical technique is demonstrated here to show the essence of social scaling.

Suppose a set of statements, I-VI, were presented to 18 respondents, A-R. Each person gave a "yes" or "no" response to each statement. The results were then tabulated (Table 3.4, with only the "yes" replies being shown for clarity). After summing the number of affirmative replies for each respondent (i.e., row totals) and for each item (i.e., column totals), the results were rearranged

Table 3.4 Responses to a Pre-Test Used for Constructing a Scale

Respondent	I	II	III	IV	V	VI	Total
A	Y	Y	Y			Y	4
B			Y				1
C		Y	Y			Y	3
D			Y	Y			2
E			Y			Y	2
F		Y	Y	Y		Y	4
G		Y	Y				2
H	Y	Y	Y		Y	Y	5
I			Y				1
J	Y	Y	Y	Y		Y	5
K	Y	Y	Y	Y	Y	Y	6
L				Y			1
M		Y	Y	Y		Y	4
N			Y			Y	2
O	Y	Y	Y			Y	4
P							0
Q		Y	Y			Y	3
R			Y			Y	2
Sum	5	10	16	6	2	12	51

(Y = Affirmative response)

according to row and column rankings of totals (Table 3.5.a). It would appear from this rearrangement that the test measures variation in attitudes with respondent K agreeing the most and respondent P the least with the set of statements. At this stage however, primary concern is with the ordering of the statements.

According to these preliminary results, Statement III is the one receiving most agreement—it is on one end of the measurement scale. Statement VI measures the next position. Everyone in the pre-test who answered "yes" to Statement VI replied the same to Statement III. Statement II is even further along the continuum as evidenced by the response pattern. Notice, however, that respondent G gave a "yes" to Statement II but not to Statement VI. An analyst would expect G to say "yes" to Statement VI if the statements were perfectly ranked since Statement II would imply Statement VI. Because this irregularity in the response pattern occurred only once, the statement should be retained. The next statement, however, produced several irregular responses. Respondents H, A, and O did not agree to Statement IV but they did reply "yes" to statements farther along the scale toward the disagreement pole. Respondents D and L said "yes" to Statement IV even though they disagreed with statements ranked more toward the agreement (yes) end of the scale. With so many irregularities associated with Statement IV it should be discarded. The rest of the responses fit the pattern desired from a set of ranked statements given a diverse population.

To produce a perfect response pattern, the replies of respondents H, A, O, D, G, and L need to be different. In other words, the set of statements does not adequately measure a third of the

Table 3.5   Responses from Table 3.4 Re-Arranged

a. First Ordering

Respondent	Statement III	VI	II	IV	I	V	Sum
K	Y	Y	Y	Y	Y	Y	6
H	Y	Y	Y		Y	Y	5
J	Y	Y	Y	Y	Y		5
A,O	Y	Y	Y		Y		4
F,M	Y	Y	Y	Y			4
C,Q	Y	Y	Y				3
E,N,R	Y	Y					2
D	Y			Y			2
G	Y		Y				2
B,I	Y						1
L				Y			1
P							0
Sum:	16	12	10	6	5	2	

b. Scaled Ordering

Respondent	Statement III	VI	II	I	V	Scale
K,H	Y	Y	Y	Y	Y	5
J,A,O	Y	Y	Y	Y		4
F,M,C,Q	Y	Y	Y			3
G	Y		Y			(3)
E,N,R	Y	Y				2
D,B,I	Y					1
L,P						0

respondents (6 of 18). This is a much higher rate of failure than is recommended by scholars working with such scales.[10] However, if Statement IV is removed, then only one response (the one for respondent G) is "out of place." The six per cent (1/18) failure that now results is less than the recommended tolerance of ten per cent, so the set of remaining statements is acceptable. Thus, the scale measures from 0 through 5 (or, any other group of six numbers), with no affirmative responses on one end and five "yes" answers on the other end (Table 3.5.b). A respondent who agrees with Statements III and VI is assigned a value of 2. Note that a future respondent who, for example, answers "yes" to Statements III, II, and I would be measured as a 4 in spite of only three affirmative replies because Statement I is ranked on the scale to imply four "yes" answers.

The development of an attitude scale is difficult. For this reason, most researchers do not attempt to construct an original one. The absence of a specially designed scale is not necessarily a limitation, however, because there are at least two alternatives. One is the existence of numerous standardized surveys developed for measuring various kinds of attitudes. One of these surveys may be close enough to the needs of a particular researcher to justify using it rather than constructing an original one that is "exactly right."[11] The alternative is to use an original questionnaire, but then utilize statistical scaling during the analytical stage of research. As a result, the researcher does not need to depend on specialized data acquired by a unique measurement scale.[12]

Evaluation and Applications

Evaluation

Because the establishment of relationships among phenomena depends on measuring their variations, measurement is an integral component of scientific methodology. In fact, the collection of data for geographic research is basically a measurement of selected characteristics of the relevant variables. Faced with the necessity of measurement, the researcher must decide the kind of measurement that is appropriate. A primary decision focuses on the level of measurement that is suitable for collecting information, analyzing the data, and solving the research problem. Among the choices of nominal, ordinal, and parametric data, no one level possesses all the advantages or disadvantages. In situations where the researcher observes objects or persons visually, measurement may appropriately be at the nominal level. In other situations measurement may be accomplished quantitatvely with the assignment of numbers to individuals. In all cases the researcher should consider the measurement procedure in terms of validity, reliability, precision, and accuracy.

Decisions about the method to be used to measure the research phenomenon and relevant variables should be made during the preparatory stage. This statement is easily accepted when the measurement procedure is expected to involve special instruments and/or tailor-made questionnaires. In contrast, when the plan for collecting information involves "only" looking at the landscape or watching a group of persons, the amateur researcher may be tempted to go to the field and start recording facts without any prior preparation. As discussed in this chapter the translation of concepts to actual data forces the field worker to make decisions about identification and classification, even if done subconsciously. The researcher who seeks high quality data, therefore, should consciously deal with these various aspects of the concept-to-data translation and decide on measurement procedures prior to going to the field.

Applications

To gain practice in this phase of research one or both of the following exercises may be tried.

1. At the end of the previous chapter it was suggested that a windshield tour to examine an area which could be subsequently measured for variations in visual attractiveness might be taken. Now design a classification scheme that can be used to differentiate degrees of attractiveness in that area. Specify the criteria for such a scheme and how they will be used to define various classes of visual attractiveness. Attempt to make an original scheme; but to get other ideas examine the classificatory schemes developed by others.[13]

2. Select a topic that is controversial (but not necessarily geographical for this practice run) and write several statements that express varying attitudes, opinions, or positions on that topic. Word the statements so persons could express their feelings about each on a Likert-type scale.

Then ask several friends (and other people if all your friends think alike) to respond to your statements. Examine the results to see how successfully the set of statements differentiated the varying viewpoints on the topic.

Notes

1. R. Abler, J. S. Adams, and P. Gould, *Spatial Organization: The Geographer's View of the World* (Englewood Cliffs, N.J.: Prentice-Hall, 1971); C. W. Churchman, "Why Measure?" in B. J. Franklin and Harold W. Osborne, eds., *Research Methods: Issues and Insights* (Belmont, Calif.: Wadsworth, 1971), pp. 129–139; D. Nachmias and C. Nachmias, *Research Methods in the Social Sciences* (New York: St. Martin's Press, 1976); J. L. Payne, *Principles of Social Science Measurement* (College Station, Texas: Lytton Publishing Co., 1975); S. S. Stevens, "On the Theory of Scales of Measurement," *Science,* 103 (1946), pp. 677–680; S. S. Stevens, "Measurement, Statistics, and the Schemapiric View," *Science,* 161 (1968), pp. 849–856; P. Taylor, *Quantitative Methods in Geography: An Introduction to Spatial Analysis* (Boston: Houghton Mifflin, 1977).
2. A clear discussion about classification along with additional references is provided by R. Abler, J. S. Adams, and P. Gould, *op. cit.*
3. Awareness of the aggregating approach is helpful because (1) it resembles the divisional approach in reverse, (2) it depends on sound field data for meaningful groups, and (3) it could be the basis for selecting criteria for a divisions classification (see Note 5).
4. *Standard Land Use Coding Manual: A Standard System for Identifying and Coding Land Use Activities* (Washington, D.C.: U.S. Govt, Printing Office, 1965). Also see: J. R. Anderson, "Land Use Classification Schemes," *Photogrammetic Engineering,* 37 (1971), pp. 379–387; R. G. Bailey, R. D. Pfister, and J. A. Henderson, "Nature of Land and Resource Classification—A Review," *Journal of Forestry,* 76 (1978), pp. 650–655.
5. In case the researcher lacks confidence in making such a decision, information about the relative importance of several attributes might be obtained by conducting a preliminary study. It would necessitate gathering data on many potential attributes, analyzing them by a clustering procedure, and then accepting the results as the most discriminatory attributes.
6. R. Likert, *A Technique for the Measurement of Attitudes* (New York: Columbia University Press, 1932); D. C. Miller, *Handbook of Research Design and Social Measurement,* 3rd ed. (New York: Longman, 1977); M. Parten, *Surveys, Polls, and Samples: Practical Procedures* (New York: Cooper Square, 1966); C. F. Schmid, "Scaling Techniques in Sociological Research," in P. F. Young, *Scientific Social Surveys and Research,* 4th ed. (Englewood Cliffs, N.J.: Prentice-Hall, 1966).
7. D. Heise, "Some Methodological Issues in Semantic Differential Research," *Psychological Bulletin,* 72 (1969), pp. 406–422; C. E. Osgood, G. J. Suci, and P. H. Tannenbaum, *The Measurement of Meaning* (Urbana: University of Illinois Press, 1957); J. G. Snider and C. E. Osgood, eds., *Semantic Differential Technique: A Source Book* (Chicago: Aldine, 1968).
8. N. H. Anderson, "Scales and Statistics: Parametric and Nonparametric," *Psychological Bulletin,* 58 (1961), pp. 305–316; S. Labovitz, "The Assignment of Numbers to Rank Order Categories," *American Sociological Review,* 35 (1970), pp. 515–524.
9. L. Guttman, "A Basis for Scaling Qualitative Data," *American Sociological Review,* 9 (1944), pp. 139–150; N. Lemon, *Attitudes and Their Measurement* (London: B. T. Batsford, 1973); L. L. Thurstone, *The Measurement of Values* (Chicago: University of Chicago Press, 1959); Note 6.
10. Some scholars consider this guide, called the "coefficient of reproducibility," to be inadequate; see L. E. Dotson and G. F. Summers, "Elaboration of Guttman Scaling Techniques," in G. F. Summers, eds., *Attitude Measurement* (Chicago: Rand McNally, 1970), pp. 203–213.
11. D. C. Miller, *op. cit.;* J. P. Robinson, J. G. Rusk, and K. B. Head, *Measures of Political Attitudes* (Ann Arbor, Mich.: Institute for Social Research, 1968); J. P. Robinson and P. R. Shaver, *Measures of Social Psychological Attitudes* (Ann Arbor, Mich.: Institute for Social Research, 1969); M. E. Shaw and J. M. Wright, *Scales for the Measurement of Attitudes* (New York: McGraw-Hill, 1967).

12. R. G. Golledge and G. Rushton, *Multidimensional Scaling: Review and Geographic Applications,* Technical Paper No. 10 (Washington, D.C.: Association of American Geographers, 1972); G. M. Maranell, ed., *Scaling: A Source for Behavioral Scientists* (Chicago: Aldine, 1974).

13. L. M. Arthur and R. S. Boster, *Measuring Scenic Beauty: A Selected Annotated Bibliography,* USDA For. Serv. Gen. Tech. Rep. RM-25 (Fort Collins, Colo.: Rocky Mt. For and Range Exp. Stn., 1976); T. C. Daniel and Ron Boster, *Measuring Landscape Esthetics: The Scenic Beauty Estimation Method,* USDA For. Serv. Res. Pap. RM-167 (Fort Collins, Colo.: Rocky Mt. For. and Range Exp. Stn., 1976).

Chapter 4

Sampling Geographic Phenomena

A person who plans to collect field data for the purpose of solving a research problem inevitably must make a decision about whether to engage in sampling or not. Usually the answer is affirmative. Then questions about sampling procedures arise. Answers to these questions must be decided before the researcher can collect data from the field. The topic of sampling logically accompanies the other preparatory decisions (Chapters 2 and 3).

Some of the questions that are answered in this chapter are the following:

In what kinds of situations will a researcher need to consider sampling?
What are the characteristics of an acceptable sample?
What are some sampling designs that are appropriate for phenomena that are studied geographically?

Defining a Population

The term *population* refers to the aggregate or totality of individuals about which inferences or generalizations are made. A *sample* is a collection of individuals selected from a population (i.e., the complete group of individuals) for the purpose of representing the population. This interdependent pair of definitions essentially states that a sample is a subset of a population which is the total set. Although it is not difficult to understand that the term population is the total set, conceptualizing and defining a particular population may not always be easy.

Consider the population of "school children." As stated this refers to all the school children in the world at the present time as well as those in all past times plus those in future times. If this is too inclusive for a particular study, the definition of the population may need to be refined to exclude those individuals not relevant to the study. Maybe the intention is to consider the set of individuals who were school children in the United States during the last two decades and who will be school children in this country during the next eight years. Even this revised definition may be too inclusive for some studies, but the decision about its relevance depends primarily on the nature of the research problem.

Defining the relevant population is associated directly with the research phenomenon and all of the variables which are hypothesized to be related to the problem variable. The specific individuals observed in a study are frequently regarded as a sample subset from a general population that is conceptualized and which must be specified by the researcher. The conceptualized population may be defined in a variety of ways depending on the way the research problem is stated. The following illustrate some of the variations in populations that might be defined for specific studies:

> all the vehicles that have ever traveled the streets of Choros;
> all the vehicles that will travel the streets of Choros during the next twenty years;
> all the rock bands that performed on the mall last summer;
> all the pilgrims that traveled to Mecca during 1975 by airplane;
> all the pilgrims that will be traveling to Mecca by airplane in 1985;
> all the kitchen sinks in the homes of Chinese families living in Canada during the early twentieth century.

Generalization of a Study

The need for defining a relevant population is also related to the research problem through the various methodological approaches. The approaches that necessitate collecting field data can be grouped according to three distinct objectives: (1) to describe the phenomena occurring within a specified area, Type I, (2) to develop general principles of location, Type II, and (3) to apply the principles of geography to a specific problem, Type III. A researcher who gathers data for a descriptive study seldom samples because the goal is to enumerate as many phenomena as possible. Most comments about sampling are not pertinent for the Type I approach. A geographer approaching a problem to achieve the third objective (the Type III approach) incorporates the concepts of sampling by commencing with the principles associated with a general population and then applying them to the problem subset. In this case, a person involved with applied geography must be certain that the specific study area is, in fact, a member of the population to which the general principles pertain. Otherwise sampling may not be required to achieve the research objectives.

The second objective, the Type II approach which generalizes the conclusions of a particular study, requires the researcher to know how the specific subset of individuals (objects, persons, events, areas) are related to the total population about which inferences are to be made. To achieve this goal, the researcher must make some sampling decisions. Because the objective of the research project is to learn about relationships that will aid in understanding locational facts in places other than the single one being examined, the particular study is always considered a sample. At a minimum, therefore, the researcher must decide whether the selected study area is truly representative of a defined (or, at least, conceptualized) population of areas.

Certainly Ms. S. had to consider initially whether her study area represented a general population. Her primary goal is to develop principles of location for educational facilities in small cities in general. One of her first decisions was to determine whether the Choros study area is representative of a defined population of small cities. She had to decide on characteristics or kinds of cities that constituted members of her total population. Because she found Choros to be representative of the population of small cities, she is confident that her results and conclusions about geographic principles can be generalized to other geographic areas. Factors involved in her decision to regard Choros as a satisfactory sample are discussed later in this chapter.

Generalization from Sampled Individuals

Regardless of the research approach taken, a researcher may rely on sampled data for another reason; namely, it is often impossible or very impractical to collect data from each individual in the study area. Rather than abandoning a research problem because data cannot be obtained about every individual, the researcher may acquire data from a few representative individuals and then generalize the results to the total population. In this context the population is normally easy to define. The major sampling task is to employ procedures that guarantee representativeness. Most of this chapter is directed to this type of sampling and generalization.

Considerations in Selecting a Sample

Information based on a sample is valuable for generalizing to a population only if the sample is *representative*. Perfect representativeness cannot be guaranteed, but various decisions can be made that will increase the likelihood that a specific sample typifies a population. Several issues that a researcher must consider involve estimating the probabilities that certain events will occur, namely, that individuals with specific characteristics will be sampled. Although an understanding of probability theory is helpful in choosing among alternative sampling designs and procedures, many decisions can be made wisely even without a formal course in statistics.[1]

One of the characteristics of a population that is easily understood is the existence of variation among individuals. If all individuals and places were identical, it would be extremely easy to obtain a representative sample because one individual would completely typify all other individuals in the population. In such an uninteresting world there would be no variation—all places would be exactly alike (except for geometric position) and all individuals would be the same. It is the variation that makes life interesting (*viva la difference!*) and lays the foundation for establishing relationships among phenomena. Variation is also the source of errors in making inferences from samples. Even a large, carefully selected sample may fail to represent some individuals in a population because they vary greatly from the more "average" ones. Therefore, sampling is concerned with both the average, usually the statistical mean, and the variation, which is commonly measured statistically by the variance.

An Initial Warning about Bias

The goal in sampling is to select a subset of individuals that accurately represents the characteristics of the total population. It is impossible to insure that this goal will be achieved, so it is necessary to design a sampling procedure that will minimize the risk of selecting a sample that is not representative. It is especially important that the sampling procedure does not produce unrepresentative samples systematically. A procedure that will repeatedly create unrepresentative samples is a *biased* procedure and the data collected from the sampled individuals will tend to contain a *bias*. Because the dangers of bias occurring in the collection of data are great, it is important that this term and its implications be clarified more. First, bias must be differentiated from a term that applies to those inaccuracies that result from chance (that is, sampling error; see below).

A simple numerical example may aid in understanding bias (Table 4.1). Assume a population of eighteen individuals, A-R, has a characteristic that can be measured by integer values. For the moment, pretend that a researcher has a magical, omniscient power that allows him or her to know the value for each individual. With this extraordinary information he/she can calculate the true population mean (which is 9). With this omniscient power one can be completely confident of the accuracy of this population fact, called a *parameter*.

If this magical power is removed and the researcher does not have the resources to measure the entire population, he/she cannot know the true population mean. Instead, the parameter must be estimated by drawing a sample of individuals.[2] The best way to obtain a representative sample is by a random procedure, which gives each individual of the population an equal chance of being chosen. A *random sample* is one that satisfies the conditions required for using probability theory, which in turn makes it possible to evaluate the likelihood that the sample is truly representative.

Suppose the sample obtained consists of the five individuals listed as Sample No. 1(Table 4.1). The mean of the five values in the sample does coincide with the population parameter (although, of course, one would not know that the sample mean was correct without having had that initial magical glimpse at the population). Note, however, that, although the population mean is represented accurately by the sample, the variability of population values is not exposed by the sample (i.e., the population values range from 2 to 13 but those of the sample vary only from 8 to 11).

Table 4.1 Four Samples from a Population of Eighteen Individuals

Population			Samples							
Individual Identification	Value		Sample No. 1		Sample No. 2		Sample No. 3		Sample No. 4	
A	8		J	8	Q	8	G	7	B	10
B	10		B	10	B	10	E	7	K	13
C	11		N	11	C	11	Q	8	G	7
D	7		Q	8	A	8	M	2	F	13
E	7		O	8	P	13	R	6	M	2
F	13			—		—		—		—
G	7		Sum	45		50		30		45
H	13		n	5		5		5		5
I	5		Mean	9		10		6		9
J	8									
K	13									
L	12									
M	2									
N	11									
O	8									
P	13									
Q	8									
R	6									

Population Sum: 162
Size of Population: 18
Population Mean: 9

The same randomizing procedure, when repeated, produces many different samples. Normally a researcher generates only one sample, but for illustrative purposes consider another sample, No. 2. The mean of these five values is 10, which does not accurately represent the population mean. The incorrect mean that occurs in the second sample is called a *sampling error,* but this kind of error is *not* critical to a person collecting data because the statistical analysis of data obtained from a random sample can include a precise statement about the probability of such errors occurring. Therefore, if data are collected according to a truly random sample, the researcher is assured that the probability of a sampling error existing in the data can be handled statistically during the analytical stage of research.

By way of contrast, assume that two other samples, No. 3 and No. 4, were selected in a biased manner (a manner which is not specified here but is discussed below). Because biased sampling procedures often produce unrepresentative samples, it may not be surprising to see an incorrect sample mean (No. 3). However, note that the other sample (No. 4) generated by the same biasing procedure resulted in a subset with a mean that agrees with the true population mean. In this artificial example with a known population means, the researcher realizes that there is no sampling error in this fourth sample. This unusual outcome, which was accidentally produced by a biased procedure, demonstrates that a sampling error does not necessarily result from a biased sampling procedure.

In the real world the population parameter is not known—to estimate it is the whole purpose of the research undertaking. Because the parameter is unknown, no one knows if there is a sampling error or not. Only the procedure for selecting a sample can be controlled. If the procedure yields a sample that is truly random, then the probability of committing a sampling error can be calculated. Of major importance, therefore, to anyone who collects data from a sample of individuals is the obligation to consider what constitutes sampling procedures that are biased and to avoid, as much as possible, those techniques.

Concern with bias and representativeness is common to many aspects of life. In a card game, for example, assurance is needed that the deck is sufficiently shuffled to guarantee that each card has an equal probability of being dealt to every hand. If one repeatedly receives poor hands while an opponent holds consistently good cards, one might reason that there are two explanations for these results. One possibility is "a run of bad luck"—a series of hands (samples) are the expected variations from the average that do occur occasionally (i.e., the sampling errors that are explained by probability theory). The other possibility is a dishonest opponent who purposely deals the poorer cards. Obviously one does not favor a biased procedure that allocates unrepresentative hands. Since the cards themselves do not reveal which of the two situations is operating, it is necessary to concentrate on being certain that the *procedure* is proper.

Politicians and other citizens are frequently curious about the probable outcome of upcoming elections, so pollsters attempt to make predictions by sampling a segment of the voting population. Great effort is exerted by professional pollsters to insure a representative sample because their reputations depend on their success in making accurate predictions. Usually their published results include a statement about the possibility of a sampling error. Also, they normally add a warning that the population being sampled may not coincide with the population which actually votes (see more under "sampling frame" below). In contrast, some office-seekers may be less concerned about the sampling procedures and the dangers of a biased sample than they are in the publication of percentages that favor their campaigns. Unfortunately, without knowing about the manner in which the data were obtained, it is difficult for the audience to evaluate the accuracy of the predictions.[3] This emphasizes again the fact that research results are a function of the methods by which data are obtained.

Bias and Non-Probability Samples

Biased procedures appear in many different forms. As a general guide, any method of drawing a sample that gives one or more individuals a greater chance of being selected than other members of the set will produce biased data. Two kinds of sampling techniques that are not based on the probabilities of random sampling are haphazard sampling and purposeful sampling.

Haphazard Selection

One procedure that does not employ a random sample is usually termed a *haphazard selection,* but may also be named accidental sampling. It is sometimes erroneously called "random" sampling, but the procedure does *not* guarantee randomness. Haphazard selection refers to choosing individuals arbitrarily without any conscious attempt by the researcher to pick particular members of the population. However, a multitude of unknown and undetected personal factors that affect human decisions invariably create a biased sample. It is virtually impossible for anyone to act totally without bias, so the selection of individuals, even haphazardly, cannot meet the requirements for a truly random sample.

The dangers in a haphazard selection of individuals is illustrated by a plan Mr. C. devised for gathering information about the location of the Choros school building. He feels that the less well-to-do families in the city should not have to spend more time and money getting their kids to a new school location than they already expend taking students to the present school site. He decided to ask families with modest incomes and with elementary-age children how much time they now spend getting pupils to school and what their travel costs are. The results would give him a basis for making comparisons with predicted expenditures for an alternate site. He knew he could not contact all the families that satisfied his population criteria of income and children so he planned to direct questions to a sample of persons. He decided to sample the first twenty-five persons satisfying the income and parental criteria he met during his routine schedule of activities such as going to and from work, stopping at his favorite tavern, and going bowling. Will this procedure produce an unbiased sample?

To answer this question, a person considers whether individuals having particular characteristics will tend to have a greater probability of being selected than others. More specifically, it is necessary to be certain that individuals having attributes that are related to the research variables are selected fairly. If the variations in the research variables are *entirely unrelated* to the variations that generate the sample, then the haphazard selection might be considered unbiased. It is the existence of relationships among the phenomena being studied and the sampling design that creates bias and causes the sample data to misrepresent the true population variation. Thus, if there were no relationship between the way families transport their children to school and their other daily activities, the sampling procedure devised by Mr. C. could conceivably yield an unbiased sample.

No one knows the actual variation that exists in the population Mr. C. wants to sample, so it is almost impossible to "prove" that his sample is biased. Common knowledge, however, suggests that it is highly suspect. The people he meets at work probably have travel patterns different from those in other jobs; and the differences undoubtedly carry over to the way they transport their children to school. The persons he meets during his leisure hours very likely enjoy similar recreational activities and hold similar attitudes about their children's activities. His daily decisions about where he will go and what he will do increase the chances that he will meet other persons with similar attitudes about expending time and travel on school children. Even the way he operationalizes his sampling may accentuate the bias. For example, when he was crossing the street on Tuesday morning, he almost "met" Mr. E., but M. C. decided that that encounter did not really count as a "meeting." This decision may have been influenced by the fact that he disliked Mr. E. and generally avoided him because of widely differing viewpoints.

These comments illustrate a few of the potential interrelationships among the sample Mr. C. selects and the data he is collecting. It would appear, therefore, that the results he obtains from his data will not accurately represent the population to which he wants to generalize his conclusions. This example is the kind of data gathering error which may occur in an even more subtle form than illustrated here and which must always be guarded against by the researcher.

A modification of haphazard selection is *quota sampling,* which is a procedure designed to reduce the unconscious bias that occurs in choosing individuals. The population to be sampled (i.e., the research or target population) is first classified according to one or more characteristics that are probably related to the problem variable. Then the percentage of the population in each class is estimated. These percentages serve as the proportion, or quota, of sampled individuals to be selected from each class. For instance, if it is estimated that 62 per cent of the target population consists of males, then the researcher attempts to choose a sample with 62 per cent males and 38 per cent females.

If a researcher were normally inclined to select females, the quota system might produce a better representation than not using it. It will not necessarily do so if male-female differences are not related to the variable being studied. If the technique for choosing individuals within each class is still done accidentally, the sample remains a haphazard selection. Furthermore, usually other characteristics besides the one(s) used to classify the population are also related to the research phenomenon, and these may affect the choice of sampled individuals. If these other variables are unknown or ignored in establishing classes and setting quotas, the procedure fails to eliminate biased sampling.

Purposeful Selection

Another type of non-probability sample is a *purposeful selection.* To obtain this kind of sample a researcher must consider all of the relevant variations in the population and then choose a few individuals that are "typical."[4] The researcher makes a judgment about what constitutes an average member, or a set of representative members, of the population and then selects individuals that meet those qualifications. Although conceptually sound, a representative sample is actually very difficult to achieve by this method.

To select a typical or average individual, a person requires enough information about the total population to be able to estimate what its average is. In many situations when the goal is to learn about the characteristics of a phenomenon, the researcher does not have enough knowledge initially to determine what is average or typical. Even if an average for one characteristic is determined, this does not insure that an individual possessing that average is typical in terms of other attributes. Furthermore, the averages do not provide information about variation. A truly representative sample should also contain some individuals with extreme characteristics. Carrying this logic a step further would require a researcher to divide the population into several classes based on variations in one or more characteristics and then select a proprotional quota from each class. If this could be accomplished, then a representative sample might result. If this much information were already known about the phenomenon, though, additional research might not even be necessary.

In summary, a random sample usually serves the goals of representativeness better than a purposeful selection that depends on guessing at what is typical. Useful information may be gained from a purposeful sample, but the researcher should be aware of its underlying assumptions and their limitations.

Selection of a Study Area

It is appropriate now to return to the topic of study area selection. As stated earlier in this chapter, when the research methodology is to develop geographic principles, questions about the population and a suitable sample must be pondered. When the study area is regarded as a sample from a general population of areas, the researcher must decide whether the selected area is a representative sample. This decision is essentially the same as selecting a purposeful sample. It requires the researcher to specify the characteristics being measured and to estimate the variations in those characteristics.

Since the methodological goal for Ms. S. is to generalize the Choros study to other cities, her problem illustrates the selectivity task. What are the characteristics of those other cities that are also members of the population? Do they have a certain size, economic base, social composition, governmental organization, form of school financing, geographic location, or other specified attributes? How much variation exists in the relevant characteristics, and how well does Choros represent that variation?

In practice the decision about the representativeness of a particular study area is usually made from two somewhat contrasting viewpoints. One viewpoint follows the normal sampling sequence. First the population is defined in terms of the relevant characteristics and their variable limits (fig. 4.1.a, Step 1). The population is composed of all the individual areas that meet the defined criteria; each one is a potential study area. After all the members of the population set are identified (e.g., all small cities), then an individual (e.g., Choros) is selected. It might seem that the individual should be selected randomly. However, because the sample size is one (i.e., the single study area), it is better to select an individual that is estimated to be among these that are most typical (fig. 4.1.a, Step 2).

The other viewpoint commences with a specific study area (e.g., Choros). Then the researcher decides what theoretical or conceptual population this area represents (fig. 4.1.b, Step 1). This decision requires the specification of the characteristics that are relevant to the research (e.g., size of city). In addition, the researcher must consider the amount of variation that will be tolerated in defining the theoretical population and in generalizing the results (fig. 4.1.b, Step 2). In terms of the Choros study, the conclusions that Ms. S. reaches should help explain the geography of

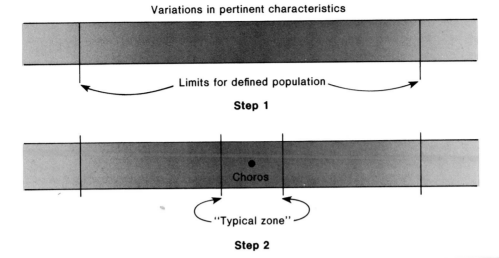

a. Selecting a study area that represents a defined
   population of areas

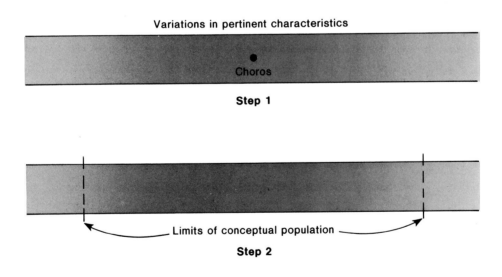

b. Defining a population of areas represented by a selected
   study area

Figure 4.1. Two approaches to selecting a study area as a purposeful sample

educational services in other small cities. What are the limits on the characteristics of other places that belong to the same population as does Choros? For example, when does the size of a city become so large that the locational relationships discovered in Choros no longer apply? She does not necessarily have to specify the limits as such, but she should define the population in terms that other scholars can use when they estimate the approximate limits of the population to which her conclusions apply.

These comments about selecting a typical study area (an individual) reinforce previous remarks about the actual delimitation of a study area. The responsibility for selecting a representative area as a sample unit from a larger population space directly affects the definition of the population. This is because the population, by definition, must be composed of all individuals, which are essentially all the potential study areas. The task of delimiting the boundaries of a study area is complicated because of the continuous nature of earth space (Chapter 1). For many nonspatial phenomena a population is composed of discrete individuals; thus, the selection of a typical element is concerned exclusively with its inherent characteristics without worrying about its extent. For a geographer the definition of what constitutes an areal unit (individual) is an added subjective decision.

Deciding on the spatial extent of an area that typifies a general population is a task facing Ms. S. Do the boundaries of the Choros school district adequately define an area in which the spatial behavior associated with educational facilities occurs? What would be the effect on her conclusions if the areal limits were restricted to walking distances to the school buildings? If the limits were moved outwards to include some of the rural community that trades in Choros, how would this definition of an individual areal unit affect the population of all such trade areas? How would each of these alternative study areas relate to, and be generalized to, the three populations similarly defined (that is, by school district, by walking zones around schools, and by small cities and their trade areas)?

In summary, before Ms. S. begins collecting data she must delimit an area with precise boundaries that she can justify in terms of her research questions and overall methodology. The proper time to make decisions about the boundaries of the study area is after the field reconnaissance (Chapter 2) but during the preparatory stage. Once the decision is made it determines the constitution of the sample individual (case) and hence the total population of areal units to which generalizations apply.

In actual practice, the decision may not appear to be as complex as described here. Often geographers choose administrative units such as cities, counties, states, or countries for their study areas because of the convenience in gathering data and ease in generalizing to other areas. However, even if the selection of a study is primarily for convenience, the careful researcher thinks about the conceptualized population represented by the study area and communicates these thoughts along with the study's conclusions.

Sample Size

The representativeness of a sample is influenced by its size. Intuitively one senses that a random sample of fifteen individuals from a population of eighteen will usually yield a mean closer to the population mean than the mean calculated from a sample size of five. Furthermore, if fifty fifteen-member samples were drawn randomly, their means would not vary from sample to sample as much as the means of fifty five-member samples would vary. Thus, it is rather obvious that sample size has an effect on how well a sample represents the total population and that the largest samples represent best.

Although a bigger sample may be expected to represent the population more accurately, it is not necessarily the better choice. A very common reason why bigger is not always better is the cost and time involved in sampling. A poll of half the votes prior to a national election might provide the data for a very good prediction; but the tremendous increase in costs over a small sample (say, 3% of the voters) is seldom justified by the slight improvement in accuracy. Most researchers find strong reasons for keeping a sample as small as possible without sacrificing too much of its representativeness. A critical question, therefore, concerns the optimum sample size that balances these conflicting objectives of increased accuracy and reduced costs.

The variety of sampling designs and the complexity of sampling theory make it difficult to give an easy answer to the question of sample size, particularly in a short text devoted primarily to other aspects of data collection. Even a basic explanation soon involves terms and ideas built on statistical theory. Nevertheless, the person who aspires to collect data from a sample of individuals needs to be familiar with some fundamental issues that face anyone, professional and amateur alike, who is deciding on the appropriate size of sample. The following comments deal primarily with variance, degree of accuracy, and confidence level.

Variance

One consideration is the variance or degree of *homogeneity* of the population. If all individuals are identical so there is perfect homogeneity (and hence, no variance), then a sample of one will represent the population. If there is moderate variation, then a larger sample is needed to capture the degree of heterogeniety. If no two individuals in the population have the same characteristics, then even a very large sample cannot represent the total population diversity. For example, when Mr. C. decided to contact only families with modest incomes and with children, he defined a population possessing less variation than presumably exists in the population of all Choros residents. Therefore, he could adequately represent this more homogeneous population with a smaller sample than would be required for a sample of all Choros citizens.

Many sampling designs attempt to gain the advantages of homogeneity by first classifying an observed, and hopefully related, phenomenon into subpopulations, called *strata*. Then each stratum is randomly sampled to form a part of the total sample. Because the goal of classification is to group individuals into classes that maximize homogeneity (Chapter 3), categorization per se creates strata each of which can be represented by a small sample. Successful stratification is dependent on acquiring data about one characteristic that is correlated with the variation in the attributes being studied.

The role of stratification is illustrated by a procedure Ms. S. is considering. She wants to conduct an in-depth interview with a sample of families having elementary-age children. One of the topics she plans to investigate is the proportion of school travel actually accomplished by the children themselves, both by walking and riding a bicycle. She decides to measure this phenomenon in miles per academic year (for example, a child that walked a half mile each morning and each afternoon for 180 days would travel 180 miles). She wants to choose a sample that represents the diversity of travel habits, ranging from those who never walked or rode a bike because they only rode in cars to those that covered a lot of miles walking or biking during the last year; yet she

Table 4.2   Distances Choros Elementary Students Walked and Rode Bicycles to School during a Year

Distance Category (miles)	Number of Students			
	Grade 4	Grade 5	Grade 6	Total
0- 25	15	5	5	25
26- 50	20	5	5	30
51- 75	20	10	5	35
76-100	15	15	10	40
101-125	15	15	10	40
126-150	5	20	10	35
151-175	5	15	15	35
176-200	5	10	20	35
Over 200	0	5	20	25
Total	100	100	100	300

does not look forward to interviewing more persons than necessary. One sampling procedure that she is considering involves stratifying the children according to grades and then randomly sampling a few from each grade. Is this a good strategy for stratifying the population of elementary students in Choros?

Although some secondary issues could affect an answer, the emphasis here is only on the merits of stratifying the population by using grade categories. The primary question then becomes, "Is there a relationship between the pupils' grade in school and their travel habits?" Stated more directly the query is, "Does a grouping of students by similarity of school class concomitantly cluster them in more homogeneous groups according to distance traveled per year?"

To illustrate the question and its answer, assume a knowledge (again as the result of temporary omniscient power) about the actual travel behavior of Choros students (Table 4.2). Also assume, for the ease of making rapid comparisons, that there are exactly 100 students that walked or rode bikes to school in each of the three elementary grades. The data are grouped by distance categories (e.g., 40 students traveled 76–100 miles during the previous year). These same facts can be presented in graphic form by histograms (fig. 4.2). Ms. S., of course, does not have all this information so one needs to empathize with her uncertainties.

If Ms S. draws a large random sample of all elementary pupils, and then graphs the data, her histogram would probably resemble the shape of the population histogram (fig. 4.2.a). However, if she reduces the sample size, there is an increased probability of getting an unrepresentative subset with a greater sampling error. To keep the sample small while also striving for representativeness, she plans to stratify the pupils by grade. Since the travel distances are related to school grade, then the histograms for the three grades will probably be similar to the three subpopulations (fig. 4.2.b). A sample from each stratum can be small yet expectedly representative because of the greater homogeneity in each class. Thus, the combined sample could remain small while exposing the variation in the total population.

Now suppose the travel distances were *not* related to school grade (even indirectly related, for example, through the factor of age). The histograms will not display distinct forms. Instead,

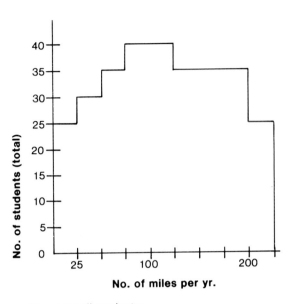

a. Travel by all students

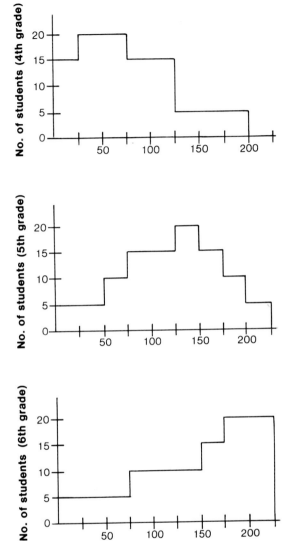

b. Travel by each grade

Figure 4.2. Frequency of travel distances by students in Choros

the histogram for each grade might be a smaller replica of the population histogram. To sample from each of these strata, therefore, would not provide any better representation than sampling from the population without stratification. For Ms. S. to stratify the population for sampling purposes would be a waste of time and effort if, to repeat, there were *no* relationship between school grade and travel distance.

Ms. S., not possessing an omniscient perspective, does not know the amount of population variation that exists in travel distances by Choros children. Only after she has already sampled the population and obtained the relevant data can she estimate the variance. Furthermore, she does not know if travel distances are correlated with school grade or not. She must make an educated guess about the probable situation. On the basis of her background understanding of the spatial behavior of school children in a setting like Choros, therefore, she justifies reducing the sample size by stratifying the population. It is this type of subjective decision that any researcher must make when considering sample size and its relationship with variance and its effective reduction through stratification.

Degree of Accuracy

A second consideration in choosing sample size is the accuracy desired. Although existence of a relationship between sampling error and sample size was presented intuitively at the beginning of this section, some implications should be clarified. The researcher may gain personal satisfaction from an enlarged sample because of the increase in expected accuracy, but this psychological benefit could be costly if the collection of data from each individual is fairly expensive and additional accuracy in unnecessary. The decision about sample size, therefore, depends on the cost of adding more individuals to a minimum sample and the necessary degree of accuracy.

Because the population parameter is unknown, there is no way to determine whether the corresponding sample statistic is accurate or not. However, by using probability theory, accuracy is measured by a value called the standard error of the statistic, which can be considered here without having it formally defined.[5] This value provides an index of accuracy that is usually expressed as either a range of values (e.g., between 82 and 88 bushels) or a deviation from the mean (e.g., plus or minus 4 percentage points from a mean of 44 per cent). Therefore, to speak of increasing the accuracy of a sample is equivalent to diminishing the standard error and thus reducing the accuracy range. The standard error is inversely related to the square root of the sample size; that is, to halve the range of accuracy requires a sample four times as large. Increasing accuracy beyond the necessary minimum is achieved at high costs.

Deciding what accuracy is the "necessary minimum" depends largely on the research problem and the utilization of its conclusions. Basically, the researcher must decide how important it is to have a narrow range of accuracy. If a national office-seeker (Mr. F.) learned from a poll that he was favored by 56 per cent of the sample votes with a possible error of 4 percentage points, most persons would declare that level of accuracy quite adequate because a majority is still assured. In contrast, if a sample poll found another candidate (Ms. G.) in a two-person contest favored by 53 per cent of the voters subject to an error of 4 percentage points, people would understand why Ms. G. wished the sample had been larger. She would have preferred to spend more money on the sample poll so the size could have been increased and the accuracy deviation decreased to, say, 2½ percentage points from 53. A third candidate (Mr. H.) ran for a non-paying position as director on the municipal library board. When the local newspaper ran a poll and learned that Mr. H. was favored by 53 per cent of the voters, subject to an error of plus or minus 4 percentage points, this news did not disturb Mr. H. He certainly had little reason to pay for another poll that would be larger and guaranteed to produce greater accuracy because the possibilities of error in the sample would not greatly affect his life. To Mr. H. the costs or effects of a sampling error were minor and did not justify the expenses of a large sample. Thus, although the degree of accuracy for a sample can always be statistically calculated and precisely stated, the ultimate decision about its adequacy rests on the researcher's judgment about the costs of basing a conclusion on a sampling error.

Confidence Level

A third factor affecting sample size concerns what is termed *level of confidence.* Obviously a person can never be absolutely certain that a sample accurately represents a population. Drawing a huge sample of 999 green beads from an urn containing 1000 beads does not guarantee that the remaining member of the population is green. Sampling involves taking chances and utilizing probability theory to reveal the amount of risk. Consequently, conclusions based on sample data should always specify the probability (e.g., 1 chance in a 1000) that an error may have occurred. Stated in a positive sense, researchers can declare numerically (e.g., the ratio of 999/1000) how confident they are that the population has been sampled accurately.

The probability of having an accurate sample increases in a mathematically precise and objective manner with the size of the sample. What is subjective is the manner in which this probability is combined with the range of accuracy. This subjectivity may be clarified by returning to the previous illustrations, each of which should have included a probability statement. In other words, the degree of accuracy is meaningful only when accompanied with a declaration of probability. Thus, the results of the first poll should state that candidate F was favored by 56 per cent of the sampled voters with, for example, a 1-in-20 chance of an error exceeding 4 percentage points. If the probability of being wrong once in twenty times disturbs the candidate, the results can be reported in an equivalent statement with a smaller chance of error occurring outside a larger range. For instance, confidence in the poll for Mr. F. might be expressed also as one chance in fifty that an error exceeds 5½ percentage points. If candidate G ran a second poll that was huge enough to narrow the accuracy range enough to keep the lower limit above 50 per cent and also big enough to provide confidence at a level of 499/500, then she might be ready to schedule an election night celebration.

Setting the level of confidence is a subjective decision which researchers must make on the basis of their willingness to accept a possible sampling error. The willingness to run the risk of a sampling error and to live with the consequences of a wrong decision varies with personalities and the seriousness of making mistakes. A probability may be placed at any level. Most researchers however choose a confidence level of 0.95 (19 in 20). If they want to be very certain of their results, they may set the level at 0.99, or even 0.999.

These three factors—homogeneity of the population, degree of accuracy, and level of confidence—must be considered when a researcher decides on how many individuals can adequately represent a population. Although they can be quantified to yield a specific sample size, the numbers are based on subjective decisions distinctive to each research project.[6] Furthermore, the tremendous variety of sampling designs makes it difficult to specify a formula for calculating sample size for each design. Even the frequently cited rule-of-thumb guidelines (of a minimum of fifty individuals per category or a 5 per cent sample) seldom satisfy the rigorous requirements of sampling theory. For major studies where the conclusions are critical and the costs of obtaining data from each individual are high, a researcher should consult a professional statistician.

On the other hand, amateur researchers should not let this question of sample size deter them from collecting credible data. For one reason, the numerical values inserted in formulas depend on subjective but informed decisions, so even formulas utilized by professionals are meaningless without the knowledge that every researcher should possess. An essential ingredient in the decision is the information each researcher has about the particular study area, the phenomenon being examined, and the purpose for the research. Another reason for not getting overly worried about calculating a precise figure for sample size concerns its importance relative to the major issue of representativeness of a sample. If the sampling procedure is not based on randomness, there is no way to specify an adequate size to insure representativeness. Therefore, the researcher's primary concern should be eliminating potential bias; the question of sample size is strictly secondary.[7]

Sampling Designs

A good sampling procedure is one that designates a few individuals that accurately represent the population being studied. Because the accuracy of a sample can never be determined without knowing the population characteristics (which are unknown), it is necessary to obtain a sample randomly so that at least the probability of accuracy can be determined. What are common procedures for obtaining a random sample? Although the main interest here is with sampling areas, a few preliminary comments about non-spatial designs will provide a foundation for areal sampling.

Unrestricted, Non-Spatial Random Sampling

An unrestricted random sample is one in which n units are obtained from a population of N units in a way that each unit has an equal and independent chance of being chosen. The common procedure for achieving a random subset of n units is by numbering the population units from 1 to N, and then drawing a set of random numbers from a collection of numbers 1 to N. "Drawing" the random numbers may be carried out by picking numbered beads from an urn (or, a reasonable imitation of this procedure), by using a Table of Random Numbers (Appendix B, Table B.1), or by generating a set of random numbers from a computer. Using a Table of Random Numbers is the most common technique because it is fast, easy and requires no special preparation or equipment. Because the numbers are already randomized, the user can select a sample set in any systematic manner such as reading down a column, up a column, or across rows.

Using random numbers appears simple enough and, indeed, it is easy to execute *if* all the members of the population can be enumerated. However, frequently each individual of a population cannot be identified and matched with a number. For example, suppose Ms. S. planned to sample "the residents of Choros." Even if she knew the exact size of the Choros population (which is a fact rarely known for any city larger than a few thousand), she would still encounter difficulty in making a one-to-one correspondence between each resident and an identification/sampling number. This task almost requires the names of all residents because there are few, if any, other ways of specifying which person corresponds to, for example, #2928.

The difficulty in identifying every member of a population becomes obvious whenever a person searches for a complete listing of all human individuals living within a study area. Telephone directories, voter registration lists, city directories, school records, and public utility files may contain the names of many residents, but none is complete. If Ms. S. were to use only the names listed in one of these sources, she would produce a biased sample because the missing names are not random omissions. Instead, the omitted persons probably possess particular characteristics, for instance, age, income, and frequency of moving, that keep them off one or more lists.

This predicament of having a large population with unidentified members may be tackled in various ways. One way, requiring perseverance, is to attempt to construct an original listing of all individuals in the population. If Ms. S. combines all the available lists of families in Choros and devotes the necessary time to field checking, she might be able to produce a fairly complete list. However, it would be surprising if Ms. S. were to be entirely successful. In cities having populations larger than Choros, this strategy is nearly impossible to implement. In fact, many of the merits of sampling are lost if a tremendous amount of time and effort is expended on building a population list.

The second strategy is by redefining the population or designating a surrogate one. If the researcher can adjust the problem variable slightly and by so doing can enumerate all the individuals, a redefinition is a feasible solution. Maybe Ms. S. isn't really interested in everyone in Choros; probably she would be satisfied with a population of persons between the ages of 6 and 95 who are not institutionalized. By redefining her targeted population in these terms, she will reduce the size of her list of citizens, including persons whose names appear on very few public lists.

When the problem phenomenon cannot be changed without greatly altering the research goal, then a surrogate population may be used. For example, if a pollster wants to predict an election, it is impossible to deal with the population of voters who will actually cast ballots, until they have done so. Instead, a different population is sampled, namely, those persons who are registered voters and who state they intend to vote on election day. The pollster substitutes the population of registered-and-intending voters for the population of actual voters. The degree to which these two populations differ, of course, affects the success of the pollster's prediction.

This second strategy is incorrectly used when a redefined or surrogate population differs greatly from the research phenomenon but this difference is ignored when results are interpreted. A candidate who believes that the population of persons who say they will vote is identical to the population of actual voters may be startled by the election results. What if Ms. S. were to redefine the population of Choros residents as the set of persons whose names appear in the telephone directory? This would be a drastic move because then she could generalize her results to only that very limited population, which in many American cities would be predominantly adult males. Thus, changing the population may change the entire research goal and the applicability of the conclusions, seldom a satisfactory way out of the predicament.

A third strategy is to use a different *sampling frame*. The frame refers to a structure for identifying and listing population units so they can be sampled. According to the original definition by Ms. S., her population consists of individuals who are residents of Choros (at a particular time). Each population unit is a human individual, and the sampling goal is to select a subset of individuals in a random manner. Because that can not be achieved, she revised her sampling frame to dwelling units, which are much easier to identify and enumerate. She hopes each dwelling unit contains one or more individuals belonging to the defined population, but she realizes a few units may be temporarily empty. Ideally, only members of the population will be revealed by this frame, but she anticipates having to determine whether the persons in each sampled dwelling unit are residents of Choros or not (e.g., just visitors). Also she hopes that every resident occupies a dwelling unit (that is, none sleeps in the bus station). Since she knows that the sampling frame does not coincide with individuals, she must decide on a procedure for dealing with those cases where multiple individuals or no individuals reside within a dwelling unit.

A sampling frame is frequently used when all the individuals of a population cannot be identified. The drawbacks to this strategy are (1) the modifications to a purely random selection of individuals, and (2) the potential confusion about the population being sampled. The first disadvantage can be illustrated by the frame adopted by Ms. S. If she assigns a number to each single-family house, apartment, and other residential unit, she can draw a random sample and each unit will have an equal and independent probability of being selected. Suppose she goes to sampled Dwelling Unit #415 and finds six persons living there, and when she checks sampled Dwelling Unit #429 she learns that it is occupied by a single person. Regardless of whether her sample includes all six individuals in #415 or only one randomly selected from the six, the probability of each individual in #415 being selected differs from that for the sole occupant of #429. Since her sampling frame consists of dwelling units rather than people, each individual in her original population does not have an equal and independent chance of being selected. In many situations the difference between a truly random sample and one with slightly modified probabilities is not very great and not too critical for employing statistics. In any case, the researcher should bear in mind this limitation.

The other drawback is the error frequently committed by amateurs; i.e., failure to distinguish between the defined target or research population and the "population" actually sampled. Thus, relying on a sampling frame may result in an unintentional biased sample whenever a researcher neglects to realize and detect the difference between these two populations. Some specific difficulties are revealed in the following two illustrations.

Mr. C. provides one illustration. Recall, he wanted to sample a population of "families with modest incomes and with elementary-age children." Fair enough; each family unit with specified income composition characteristics is an individual (in the statistical sense). Such a set of family units exists and the total population can be conceptualized. However, to identify every one of the family units in this population is a Herculean task. It is no wonder that Mr. C. wanted to draw a sample without enumerating each of the population units. But note that he sampled individual persons—not individual families. In addition to the biased sampling procedure (discussed above), he runs the danger of confusing the sampling frame with the population units. His technique does not guarantee that each family unit will have an equal probability of being selected for his sample.

The other illustration lies in the task facing Ms. S. when she wanted to sample Choros residents. As noted earlier, the list of names in the Choros telephone directory probably differs greatly from the set of residents in the city. However, this does not mean that the telephone listings cannot be used as a sampling frame. If Ms. S. assumed that every resident in Choros can be identified with one particular telephone, then telephone numbers could be used as a sampling frame. This would not resolve all the complications such as gaps caused by unlisted phone numbers and the slightly unequal probabilities of being selected because of the varying number of persons per phone. Nevertheless, it is a frame that can be randomly sampled and is closely related to the target population. Using the telephone listings as a sampling frame could be a satisfactory sampling procedure for Ms. S.

Sampling Spatially Continuous Phenomena

In geographic studies individuals frequently are places (locations) and the population is the totality of earth space within the study area. The task of sampling is complicated somewhat by this two-dimensional aspect of space. An unrestricted random sample of a continuous phenomenon differs somewhat from the procedures commonly employed for non-spatial populations.

It is impossible to specify the infinity of points in a plane, but locational variations can be approximated by an areal frame consisting of an orthogonal grid. A sample of a geographically continuous phenomenon normally commences, therefore, by superimposing a rectangular grid on a map of the study area. The intersections of grid lines define the sampling frame. Consequently, the density of lines in the grid should be chosen to achieve an appropriate degree of areal precision. A grid with a high density of lines will identify a larger number of positions, and hence more detailed spatial variation, than a grid formed by a few lines per unit of area.

Randomly Located Points

There are several procedures for randomly selecting a sample of points from the total set of intersection points. One procedure begins by drawing random numbers. Each pair of numbers designates a sample point by the intersection of the corresponding numbered grid lines. For example, a researcher might obtain random two-digit numbers by reading down the 36th and 37th columns of the Table of Random Numbers (Appendix B, Table B.1). The first two numbers are 15 and 45. That pair of numbers forms the coordinates of a position on the grid (fig. 4.3). The next two sample points should be located at 59,16 and 68,79. However, if the shape or size of the area is such that points (e.g., the one at position 68,79) fall outside the study area, they should be ignored. The procedure continues until the previously prescribed number of sample points are located on the map. The corresponding positions on the earth constitute the set of sample places where the continuous phenomenon is to be measured.

The assigning of the 2n numbers to n positions can be operationalized in ways different from that demonstrated here. Any rule for drawing random numbers and assigning them to coordinated points is satisfactory, provided it is objective (systematic) and each intersection has an equal and independent probability of being selected. One alternative procedure that some researchers have used commences by drawing r random numbers that are designated as X values and s random numbers specified as Y values. These are then paired in rs ways to produce a set of aligned points in a pattern that resembles "systematically aligned" points (see below).

By definition, randomly located points are not positioned in a uniform arrangement.[8] This means that the pattern of points will not cover all the study area equally well, and some parts will be poorly sampled. Even though sampling theory allows a researcher to state the probability that a sample represents the population poorly, it still may be disconcerting to the researcher who is aware that an unusual characteristic is missed by the uneven scattering of sample points.

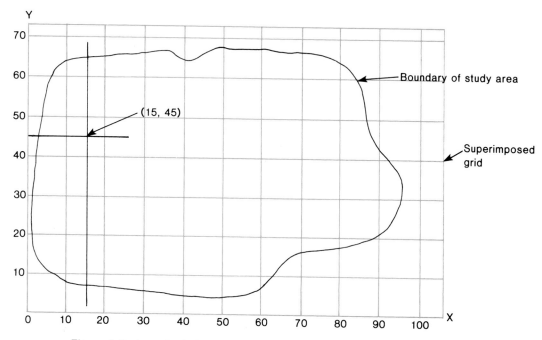

Figure 4.3. A randomly located point, positioned by a pair of random numbers matched with the coordinates of an orthogonal grid

Locationally Stratified Random Points

To insure that the sample points cover most of the study area, the researcher should first subdivide it into what are essentially areal strata. Then each areal stratum is sampled with randomly located points. According to most strategies, each stratum is sampled independently. The appropriate random numbers are those matching the grid fitting each stratum (i.e., not a grid for the entire study area).

When little is known about the areal distribution of the phenomenon being sampled, researchers often stratify their study areas geometrically. A pattern of square strata is especially easy to combine with the superimposed orthogonal grid (fig. 4.4). When strata are geometrically regular, it is possible to modify the procedure for sampling within each stratum by repeating the same random pattern of points in all strata. However, this more systematic version tends to deviate somewhat from a strict definition of randomness.

Another way of stratifying a study area is by regionalizing a phenomenon selected because its spatial distribution is known and closely related to the problem phenomenon. By regionalizing the study area and regarding each region as a sampling stratum, the researcher can achieve the same reduction in sample size for a given level of variance as accomplished in non-geographic stratification. Although a relationship between the two variables must exist (as illustrated above for school grades and travel distances) and the task of regionalization (i.e., areal classification) may be difficult, areal stratification is an important aid in geographic sampling.

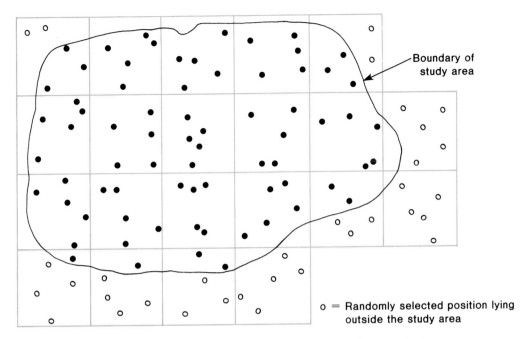

Figure 4.4. Locationally stratified random points, randomly located within each geometric stratum

As an illustration of sampling by areal stratification, look at the way Ms. S. produced sample points in Choros where she could measure noise levels. She reasoned that noise levels would vary with types of land use (e.g., areas of high noise with areas of transportation). Therefore, if she regionalized the city on the basis of a few land use types, these regions could serve as areal strata. She was able to obtain a map from the city planner that showed the locations of various types of land use in the city. On the basis of this information, she was able to divide the study area into four land use regions: residences, transportation, manufacturing, and all others. Then she randomly sampled each of the four regions or areal strata. If noise levels in Choros are associated spatially with land use types, she should have been able to detect most of the variance in the total population through the limited number of sample points. If she had used the technique of randomly locating the sample point positions without first stratifying the city, the sample points might have missed the small but critical areas expected to have high noise levels.

Systematically Aligned Points

The procedure for selecting randomly located points, either without using strata or within each stratum of subdivided study area, is somewhat tedious by the time each position is correctly placed on a base map. The task of finding the corresponding locations in the field may be even more difficult and tiresome. An alternative procedure that simplifies the selecting and locating of positions commences by subdividing the study area into n square areal strata. A random point is obtained for one subdivision, then its position within that stratum is repeated for all other strata.

This method creates a uniform pattern of points throughout the study area with the points aligned in the same manner as the rectangular grid. Following this procedure, it is extremely easy to plot the set of equally-spaced points on the base map. The uniformity makes finding locations in the field easier than searching for the scattered positions selected by the two procedures described above.

The simplicity of this method is achieved by the potential sacrifice of representativeness because the scheme produces a set of points that does not entirely satisfy the independence requirements of randomness. Furthermore, there is danger that a biased sample will result if spatial regularity occurs in the study area. Phenomena having a tendency to be spaced at regular intervals are such human features as houses in a city, farmsteads in a county, county seats within a state, and roads at almost any scale. Consequently, regularly spaced points produced by this third technique of drawing positions may consistently sample one kind of feature or characteristic and miss others. The danger of a biased sample makes this an unwise sampling system for many situations.

Stratified Systematic Unaligned Points

The fourth procedure described here attempts to combine some of the strengths of the previous schemes. It is achieved by first subdividing the study area into n square-shaped strata of equal size. One point per stratum is chosen by selecting randomly an X value for each "row" of strata and a Y value for each "column" of strata. The set of random numbers pertain to the grid fitting a stratum. If X_1 indicates the random number used for the first row of strata, Y_2 the second column, etc., then X_1, Y_2 are the coordinates of the point in the stratum located in the top row and the second column to the right. If, for example, $X_1 = 2$ and $Y_2 = 7$, then the position of the point in that stratum is 2,7 (fig. 4.5).

The final pattern of points provides extensive areal coverage yet the points are not aligned. The procedure for selecting and mapping the points is fairly easy, although persons doing this for the first time must be careful that they do not reverse the correspondence of X and Y value with rows and columns and thus end up with aligned points. Of the four geometric schemes, this last one is advocated for most geographic problems that require sampling an areally continuous phenomenon.[9] When areal stratification can be based on a regionalized and related phenomenon, that technique is frequently preferred.

Sampling Spatially Discrete Phenomena

Many phenomena occur spatially as discrete features, e.g., school buildings, dwelling units, human beings. If each individual can be numbered and if location is not important, then an unrestricted random sample will be satisfactory. If, however, location is relevant to the study, then different sampling procedures must be designed. When the individuals can be enumerated, they can be randomly selected from each of several areal strata, delimited by whatever aspect of location is deemed important. Importance might be given to uniformity of coverage, association with various land use regions, or distance from a school building.

When location is relevant and the individuals are too numerous to count, a common sampling design uses an areal frame called a *quadrat*. A quadrat is a small areal unit, often square in shape,

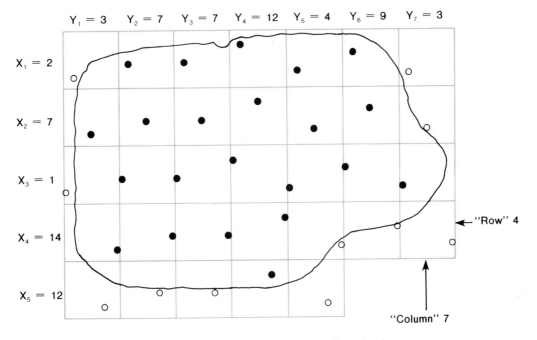

Figure 4.5. Stratified systematic unaligned points

which is superimposed on the study area at different locations (fig. 4.6). The individuals enclosed by the quadrat are then regarded like members of any other sampling frame, for example, like persons in a dwelling unit. When using this procedure, the researcher has to decide on quadrat size and on the placement of the quadrats within the study area.

The size of the quadrat depends on the density of discrete elements per quadrat and the geographic scale. Botanists, who have sampled the density of plants within an area, have created guides for choosing quadrat size, but these have limited utility for social phenomena.[10] For most human phenomena the researcher should avoid having too many empty quadrats and too many with a high percentage of the total number of individuals. The geographic scale refers primarily to whether the sampling is done directly in the field or first by map. If the quadrat is used in the field, it might be a square wooden frame that is placed on the ground, for instance, to count the number of discarded gum wrappers within a square meter on the school playground. Otherwise, the quadrats may be drawn on a base map of the study area, and later their earth-area equivalency is observed in the field.

Superimposing the quadrats on a map involves selecting the locations and, if the quadrats are not circular, their orientations. The locations can be obtained by a set of randomly located points by any of the techniques discussed above, with each point being the center, corner, or other predesignated position in the quadrat. For example, the quadrats in Figure 4.6 were located according to the coordinates of the southwest corner of each quadrat's originating position. The orientation was determined by drawing another random number (in addition to the two coordinates) that specifies the amount of rotation (e.g., by degrees 000 to 359). The enlarged quadrat

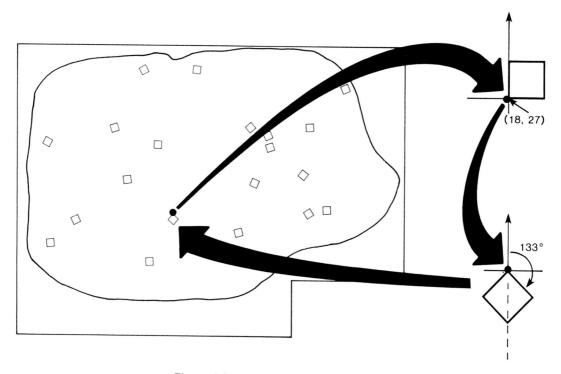

Figure 4.6. A randomly located quadrat

in Figure 4.6 was rotated clockwise from the original orientation for 133 degrees because 133 was the three-digit number selected randomly for this quadrat.

Randomly selecting small areas is a technique utilized by many social scientists who are not interested in area or location per se but who need a practical sampling frame. Stated otherwise, when the individuals of a population cannot be enumerated, often a researcher will use a geographic frame even when location is not a pertinent variable. Frequently scholars, needing a sampling of persons living in a metropolis, cannot even construct complete lists of dwelling units, let alone create population lists of individual persons. As an alternative they use areal units such as city blocks or other quadrats as a frame. In evaluating this kind of sampling frame, the researcher must estimate the degree of coincidence between the population actually sampled, which is the totality of all areal units, and the target population, namely, the set of individuals about which conclusions are stated. Very few phenomena are uniformly arranged over earth space, so the probability of an individual in the population being selected for the sample is rarely the same as the probability of a randomly placed quadrat encompassing that individual. Again, the reader should heed the warning: failure to distinguish between the target population and the population being selected by a sampling frame may conceal a serious bias.

Other Designs

   Phenomena can be sampled according to their geographic location by several other designs.[11] Three schemes that are not necessarily mutually exclusive of each other but which demonstrate other aspects of sampling that the researcher should consider are cluster samples, traverses, and multistage samples.

Cluster Samples

   A cluster sample refers to the practice of obtaining several samples in close proximity. The reason for such a deliberate spacing is not because of a spatial problem per se but because of convenience in measuring several samples without traveling long distances between them. The procedure for producing a cluster sample is to divide the study area geographically, to select a sample of those areal subdivisions, and then to sample the phenomenon within the selected areas. Although this initial description appears to resemble stratified sampling, it differs in the objective for dividing the study area. The objective in stratified sampling is to minimize variation within each sampled area; in cluster sampling the objective is to retain the population heterogeneity.

   To illustrate the difference, consider a hypothetical population of twenty-five areal units that can be categorized according to one criterion into five classes, e.g., A, B, C, D, and E. Those twenty-five individuals are spatially distributed as shown in Figure 4.7.a, but the mapped information is not known. The task is to ascertain the information through a sample of five units. Divide the study area into five areal subdivisions and select a sample in such a way that the sample set represents the total variation in the population. The spatial distribution of the elements makes it possible to accomplish an ideal division of the area according to these two contrasting objectives. For sampling by areal stratification, divide the study area in a manner that minimizes the variation within each stratum and maximizes the differences between strata. The solution is Figure 4.7.b. This kind of subdivision makes it possible to select one sample from each stratum, a strategy which produces a desired sample of five and represents the total variation in the population.

   For the other objective, which leads to a cluster sample, delineate five areas such that each division represents maximum variation within itself. The purpose is to be able to obtain all the samples entirely within one areal subdivision where the individuals are clustered close together. The geographic solution is shown by Figure 4.7.c. In this idealized example the five subdivisions are identical; hence, a cluster sample from any one of them represents perfectly the total variation in the population. Cluster samples in the real world are seldom as efficient in representing the population as illustrated in this hypothetical example, mainly because variations in most phenomena tend to occur in somewhat homogeneous groups, districts, or regions. It is easier to regionalize a study area for the purpose of sampling by areal stratification than to form the heterogeneous subdivisions required for efficient sampling in clusters.

Traverses

   Traverses, which were popular with geographers in the past for sampling areally continuous data, consist of lines across a study area along which data can be collected.[12] In this respect the procedure resembles a series of aligned points which are so numerous that they form a complete line. In fact, it is difficult to speak of sample size or know how many sampling "units" really occur along a continuous line. In any case, traverses require more sampling than areal sampling

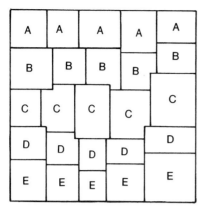

a. Distribution of individuals / areal units

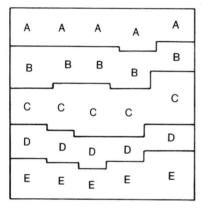

b. Division by areal stratification

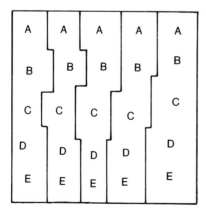

c. Division for a cluster sample

Figure 4.7. Dividing an area for a stratified and cluster sample

by random points. Generally the traverse as a sampling procedure for collecting data is less common than other methods.

Sample size and similar statistical questions are not issues if traverse locations and directions are not selected by a randomizing procedure. If traverses are placed purposefully, then their representativeness cannot be evaluated statistically but must be judged like other purposeful samples (see above). As indicated earlier, traverses can be useful for making preliminary observations during a field reconnaissance. Also, they can serve as guides for collecting data that will be subsequently shown as profiles of spatial change.

Multistage Samples

Multistage sampling refers to sampling by a sequence of steps or stages. Often it is difficult to design a sampling procedure that is both representative and small, unless sampling is done in several steps. Cluster sampling, in which the set of subdivisions is sampled and individuals drawn from the selected areas, constitutes a two-stage sample. This same kind of procedure can be extended to several stages.

As an illustration, consider the procedures designed by Mr. M. for sampling U.S. families having children in school. His research hypotheses dealt with travel behavior and included the variables of income, density of population, and location within the country. For stage one he stratified the U.S. areally by the nine major census regions. Next he stratified the counties in each region into five classes according to median family income, as reported by census data. His first sampling was a random selection of four counties from each of the 45 strata.

The second stage required stratifying each of the 180 counties into three kinds of areas: urban, town, and rural. Mr. M. realized that not all three strata would occur in every county and that the areal strata might be discontinuous, especially the several residential clusters classified as town. He sampled the rural areal stratum in each county by randomly placed quadrats. All individuals, which were families with school children within the quadrats became a part of the sample. For the town stratum in each county he regarded each areally separate part as a substratum. For example, if a particular county has a ring of dense settlement around a city, that dense ring constituted one areal unit. If, in addition, the same county had four distinct villages, he would sample from each of these four areal units. Again, the sampling within each of these substrata was achieved by randomly located quadrats.

The areas classified as city strata were sampled through a more complex procedure that involved another stage. Mr. M. areally stratified each city according to median income per census tract, then sampled each stratum (i.e., income region) by using quadrats. However, he did not include all the individuals (qualified families) residing in each quadrat. Instead, he performed another sampling stage in the field. This field task consisted of identifying and numbering all dwelling units, drawing an unrestricted random sample, and including the selected individual units as part of his overall sample. He assumed that the final sample represented the diversity of travel behavior in different parts of the United States by families living in areas typified by various incomes and densities of residences.

Evaluation and Applications

Evaluation

An inventory of techniques used in collecting data invariably includes the topic of sampling because many research conclusions can be based only on the limited number of individuals actually observed but which in turn represent a larger set of individuals, i.e., the population. The degree to which general populations are accurately represented by the selected individuals depends on the effectiveness of sampling. To evaluate the utility and effectiveness of sampling, consider two questions: (1) Why sample? and, (2) What is an effective sample? These are essentially the same questions all researchers must answer for their own projects during the preparatory stage of research.

Reasons for sampling or not sampling are given here at two scales or levels of research. One level involves the methodological approach and views the study area as one spatial individual from a population of potential study areas. At this level, the role of sampling varies from never to always. The study area is never considered a sample in a descriptive study (Type I) when the purpose is to describe the locational facts within a particular area. The study area must always be conceptualized as a sample when the researcher attempts to generalize the spatial relationships observed in the study area to a population of areas (by the Type II approach). In between is the methodological approach of applied geography, which resembles that of descriptive geography because it too commences with a study area already specified so there is no question about selecting a sample area. Applied geography differs from the descriptive approach because the researcher attempts to solve a problem in a specific area by applying the geographic knowledge acquired from a population of other areas. In this sense, the study area might be considered a sample from a population of places.

The question of whether to sample or not also arises at the level of research that pertains to sampling within the study area. At this scale the question focuses on whether the researcher wants to observe all individuals within the study area or only a subset of them. Reasons for and against conducting a sample are often expressed in terms of the practicality of specific procedures.

In a few rare cases the act of measurement may alter an individual so much that measuring the total population almost destroys it. For instance, if plants must be brought into a laboratory for a complete set of measurements, it would destroy the entire vegetational cover to measure the population. It is better to sample the population and sacrifice only a few plants.

In most cases the reasons for sampling are related to the expense and time required to measure each individual in a large population. Even such basic information as that obtained by a national census is collected in the United States only once in a decade, partly because of the tremendous expense and difficulties involved with contacting every person in the country. Most researchers do not have the financial backing to measure all individuals for all the variables in a project. Also, in many research projects individuals need to be observed at approximately the same time, but it is usually impossible for one researcher to measure all individuals during a short period of time. In such a situation, the researcher must either hire many field workers for a short time or sample the population. Most choose the latter.

For many data needs, samples can provide very satisfactory information. Even researchers collecting data for resource inventories and similar descriptive studies usually sample areally continuous phenomena (e.g., soils). Certainly samples are common to many aspects of our everyday activities. For most geographic problems, researchers find that benefits accruing from the data acquired from samples far outweigh the risk that a sampling error will occur.

The probability that a sampling error has occurred in a random sample can be calculated statistically. Therefore, this factor of uncertainty can be accounted for by making the sampling procedure truly random. Concern about an effective sample, therefore, should concentrate on other factors that may produce erroneous results. These can be summarized under the broad statement: the sample must represent the population. The degree to which a particular procedure will produce a representative sample depends partly on the homogeneity/variance in the population and the way the procedure is designed to detect that variance. But a factor of greater importance, and one that all researchers must guard against when designing sampling procedures, is bias. A sample is not representative if there is a greater probability that individuals with certain characteristics related to the population phenomenon will be selected than certain other individuals with different characteristics. Bias is such a critical element in sampling procedures that it becomes an important criterion for evaluating the specific field techniques that are presented in the remaining chapters.

Applications

Examples of specific sampling designs are common to many articles about research in the geographic literature. An article that describes quite completely a specific multistage sampling procedure is one written by W. F. Wood in 1955.[13]

For some personal experience with sampling, try one or more of the following problems.

1. Listed below are two populations, each with a possible sampling frame. For each pair describe conditions that (a) would probably not produce a bias, and (b) probably would produce a biased sample.

Population	*Frame*
Adults in a city who watch TV	The first half of the city telephone directory
Persons who go to National Park X during the May–September season	Passengers in all vehicles who go through the entrance of Park X during the first week in June

2. Figure 4.8 shows the spatial distribution of an areally continuous phenomena and the proportional amount of total area in each of its five classes. Make the realistic assumption that you do not possess this map but that the depicted phenomenon does exist in the field in the pattern shown here. If you select a sample point and go to the corresponding field position, you will observe the class that is shown on the map. Make four copies of the map, then sample the area according to each of the four procedures described for geometric sampling of spatially continuous phenomena. Compare the results of the four samples in terms of their success in representing the actual proportion of area in each of the five classes.[14]

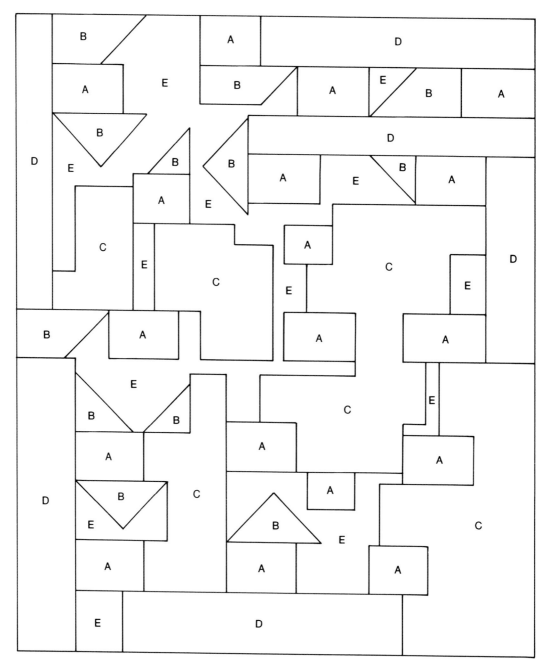

Figure 4.8. The spatial distribution of five classes of an areally continuous phenomenon

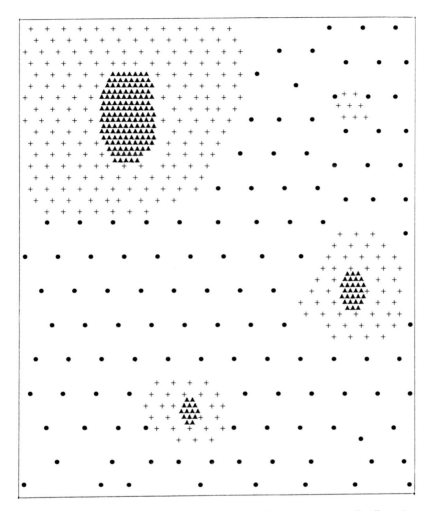

Figure 4.9. The spatial distribution of three classes of an areally discrete phenomenon

3. Figure 4.9 shows the spatial distribution of a geographically discrete phenomenon, which is differentiated into three classes: •, ▲, and +. Assume that, although you know the approximate location of most of the individuals, you do not have the data about their classes. Design a sampling procedure for estimating the percentage of individuals of each class.[15]

Notes

1. Almost any text dealing with basic statistics would be helpful. Ones pertaining to geographic problems include: R. Hammond and P. McCullagh, *Quantitative Techniques in Geography: An Introduction* (London: Oxford University Press, 1974); and P. Taylor, *Quantitative Methods in Geography: An Introduction to Spatial Analysis* (Boston: Houghton Mifflin, 1977).

2. A sample of 5 from a population of 18 is too small for statistical reliability, but the number is kept small in the illustration to simplify the arithmetic.

3. For ways social surveys can be manipulated to produce the results a pollster wants, see M. Kinsley, "The Art of Polling," *The New Republic,* 184 (1981), pp. 16–19; also see "Wording of Response Choices," in Chapter 6.

4. The term "purposive sample" sometimes is used to refer to choosing those persons who possess more or better information than the average person, but such usage negates the meaning of a sample.

5. The reader familiar with the concept of standard error may want to look at the contents of Note 6. Those who are unfamiliar with this basic statistical term should either accept it as an undefined term or consult one of the books listed in Notes 1 and 11.

6. For the mean, sample size, n, is calculated by $\frac{\sigma^2 z^2}{T^2}$, where σ^2 is the estimated variance, z is the confidence level for a normal distribution, and T is the tolerated deviation from the mean. For a percentage, n is obtained by $\frac{P(100\text{-}P)z^2}{T^2}$ where P is the estimated population percentage.

7. It might be argued that a third reason is that sample size is more critical to the analysis and interpretation of data than to the collection of data per se and, therefore, is not appropriate for this text; yet, the interdependence of data collection and data analysis is great enough that this contention is rather weak.

8. Patterns formed by a set of points are usually classified as random, more clustered than random, or more uniform than random. Accordingly, a random pattern is neither uniform nor clustered, although occasionally a randomly generated pattern of points may appear somewhat uniform or clustered.

9. B. J. L. Berry and A. M. Baker, "Geographic Sampling," in B. J. L. Berry and D. F. Marble, eds., *Spatial Analysis: A Reader in Statistical Geography* (Englewood Cliffs, N.J.: Prentice-Hall, 1968), pp. 91–100.

10. P. Greig-Smith, *Quantitative Plant Ecology* (London: Butterworths, 1964); K. A. Kershaw, *Quantitative and Dynamic Ecology* (London: Edward Arnold, 1964); and R. L. Smith, *Ecology and Field Biology* (New York: Harper & Row, 1966).

11. J. M. Blaut, "Microgeographic Sampling, A Quantitative Approach to Regional Geography," *Economic Geography,* 35 (1959), pp. 79–88; W. G. Cochran, *Sampling Techniques,* 3rd ed. (New York: Wiley, 1963); J. Holmes, "Problems in Locational Sampling," *Annals, Association of American Geographers,* 57 (1967), pp. 757–780; N. C. Matalas, "Geographic Sampling," *Geographic Review,* 53 (1963), pp. 606–608; M. Parten, *Surveys, Polls, and Samples: Practical Procedures* (New York: Cooper Square, 1966); R. J. Pryor, "Sampling Frame for the Rural-Urban Fringe," *The Professional Geographer,* 20 (1968), pp. 257–261; H. B. Rodgers, "Random Sampling Techniques in Social Geography," *Geographia Polonica,* 18 (1970), pp. 139–156; P. Taylor, *op. cit.;* and F. Yates, *Sampling Methods for Census and Surveys,* 3rd ed. (New York: Hafner, 1960).

12. P. Haggett and C. Board, "Rotational and Parallel Transverses in the Rapid Integration of Geographical Areas,"*Annals, Association of American Geographers,* 54 (1964), pp. 406–410; G. E. Matzke, *Line Census Techniques: An Application in Biogeography,* Discussion Paper Series, No. 11 (Syracuse, N.Y.: Department of Geography, 1976); and M. J. Proudfoot, "Sampling with Transverse Traverse Lines," *Journal of the American Statistical Association,* 37 (1942), pp. 265–270.

13. W. F. Wood, "The Use of Stratified Random Samples in a Land Use Study," *Annals, Association of American Geographers,* 45 (1955), pp. 350–367; but note the unusual constraint that requires the sample to be distributed within three specified areas, p. 353.

14. The true proportions, given in percentages, are: A = 16½, B = 8, C = 28½, D = 22½, and E = 24½.

15. The exact percentages are: • = 21.24, ▲ = 32.21, and + = 46.55.

Chapter 5

Collecting Data by Observing

Direct observation is "the archetypical technique of scientific inquiry in virtually every field."[1] This is especially true in geography where fieldwork has traditionally meant traveling to the study area for the purpose of observing one or more phenomena. While directly observing the relevant phenomena, the researcher makes decisions about what is being seen and then records those decisions as measurements. It is this general observational method of collecting data that constitutes the core of this chapter.

Various aspects of this data-collecting method are discussed by considering the following general questions:

How can a classification scheme be utilized to obtain useful field data?

What techniques can be employed to collect information about inanimate objects?

How can data be obtained by observing human subjects?

How do subject awareness and research control affect various observational techniques?

What factors should a researcher consider when evaluating the comparative advantages of different techniques?

The term "observation" needs clarification because sometimes it refers to sensing in general and sometimes only to seeing. By observation some scholars use the term to mean "any sensory perception not only visual, of external cues which help us to understand" phenomena.[2] Other scholars often separate observational methods from interviewing and similar techniques that depend more on verbal communication than on visually sensing phenomena. Confusion about the interpretation of the word is not entirely resolved here. In this text, observation usually refers to receiving stimuli from phenomena by any of the human senses. In this chapter, however, the emphasis is mostly on techniques that sense phenomena visually.

Visual observations are basic to gathering data. This is mainly because people depend so much on visual stimuli for knowledge. No mechanical instrument is so effective as the human eye and brain in scanning a scene, extracting the relevant phenomena, and comprehending relationships among objects. This mental processing occurs almost simultaneously. The mind can perceive and encode information as rapidly and continuously as the environment changes. The fact that visual observations are such a common way of gaining information means that everyone has developed some proficiency in this method of gathering data.[3] This accumulated experience can be an asset for the beginning researcher who has insufficient time to learn more esoteric methods. This high dependence on visual observations can also be a disadvantage when objective measurements are needed. Because visual observations are so common, the process of extracting data is unconsciously accepted with very little contemplation about the nature of the information input. As discussed previously (Chapter 3), the mental procedures for accomplishing nominal measurements are frequently done without an awareness of the decisions being made. Adherence to scientific methodology, however, demands that data be acquired by techniques that are as objective as possible. This requirement applies not only to complex procedures that force intellectual concentration but equally to those that involve "merely observing". Therefore, it is important that the beginning researcher gain experience in observing phenomena critically because the ability to make observations is the *sine qua non* of any scientist.[4]

Observational Techniques

Discussion about techniques is complicated by the fact that geography involves a wide variety of phenomena, rather than a single kind of object or event as is the case in the more topical disciplines. For example, methods for obtaining data about a community of persons may not be appropriate for gaining information about a set of inanimate buildings. Therefore, to simplify the applicability of specific techniques, the general method of visual observation is summarized here in four main categories according to variations in the phenomena being observed and the degree of researcher control: (1) observing inanimate objects, (2) observing human subjects unobtrusively, (3) observing human subjects as a participant, and (4) observing human subjects within controlled conditions.

Observing Inanimate Objects

Many characteristics of the natural and manmade environment (NV and CV of Chapter 1) are normally observed and measured visually. Although there are various specialized instruments designed to measure natural phenomena quantitatively (but which are omitted from this text), many measurements relating to inanimate objects rely primarily on the qualitative measurements of visual observations. As discussed in Chapter 3, many of these qualitative measurements are the nominal and ordinal ones achieved through the use of classification schemes. Thus, the main emphasis in this section is on the task of objectively classifying features in the field.

Various aspects of field classification can be illustrated by looking at the work of Ms. S. She knew that some Choros citizens opposed constructing a new school building in the southeast part of the city because they believed such a location would provide unfair benefits to the richer sector. She decided to estimate the areal extent of such a "richer sector" by mapping the region of affluence. She was not able to collect data about the income of each family directly, but she reasoned that the value of the residence would indicate the wealth of the occupant family. No data on house values were available to the public, so she decided to collect such information in the field by estimating residential values according to building appearance. Because she knew this would be a big task, she limited the mapping to the southeast quadrant of the city and persuaded two college friends to help her.

The measurement task facing Ms. S. involved (1) identifying a residential unit, (2) finding its position in a locational system, and (3) classifying it according to its economic value. The first two tasks were not too difficult. Identification of a residence was fairly easy after she wrote an operational definition, which involved decisional procedures similar to those employed for identifying vehicles. Likewise, finding the location of each identified residence was not complicated. She had a large-scale base map that showed all the city streets. She enlarged the map and cut it

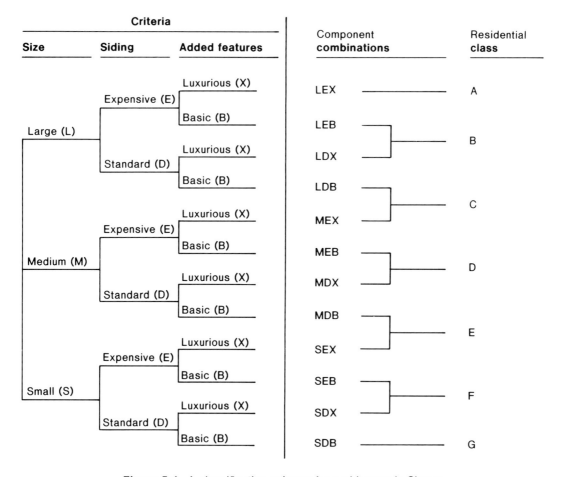

Figure 5.1. A classification scheme for residences in Choros

into clipboard-size sheets for use in the field. By finding the correspondence between the map and street positions and by doing some pacing along the city streets, she was able to indicate residential locations with adequate areal precision.

Field classification was more difficult. Her objective was to assign one of seven classes to each residence according to the criterion of economic value. Her strategy was to estimate the value by examining three components visible from the street. One criterion was size, which she subdivided into three classes (large, medium, and small) based on linear dimensions of the structure. Another component, which she called quality of siding, was divided into two categories (expensive and standard). For each of these two classes she listed the materials (e.g., brick, wood, composition board) that would help make the assignment task more objective. The third component, dealing with the existence of additional features such as a garage or swimming pool, was divided into two classes (basic and luxurious). After categorizing these three separate components, she combined them in an objective manner that assigned a measurement (class) to each residence (fig. 5.1).

Her classification scheme should not be regarded as a model to be used elsewhere. Frankly, it is rather simplistic and not too plausible, but these less desirable qualities may allow attention to be focused on other aspects of the classificatory task. The main concern here is not with the design of a classification scheme itself, but rather with procedures for operationalizing it in the field.

Classifying Objects

To classify objects in the field the researcher must make a decision similar to the one described in Chapter 3 for identifying an individual, namely, a decision about the belongingness of the object to a class. This involves matching a particular individual with a class name or numeral. By assigning a numeral, the researcher measures the individual and, thus, obtains a fact that contributes to the set of data. This measurement decision is made by the researcher at the moment each individual is observed under the varying conditions of the field environment and the researcher's personal well-being. The classificatory decision about each individual is made typically by a field worker only once with, unfortunately, a limited number of checks on consistency. For these reasons, the conscientious researcher should attempt to make the decision about the classificatory assignment for each individual as objective as possible.

The most common way of increasing objectivity and consistency in decisions is to write a thorough description of the classification scheme. A complete description includes the criteria used to divide the population into classes and the characteristics that distinguish each class. Descriptions are limited by the same factors that restrict any verbal communication, so major terms require accompanying operational definitions. However, definitions must ultimately depend on words that are called *primitive terms,* which means the terms are assumed to be understood because they cannot be defined more specifically.

This last point might be clarified by referring to the classification scheme used by Ms. S. One of the components in her scheme dealt with quality of house siding, which she divided into two classes. To be consistent in her own judgment of quality and also to communicate her intended division between "expensive quality" and "standard quality" to her two helpers, she listed the various kinds of siding that belong to each category. One of the siding terms was composition board. But, what is meant by composition board? Shouldn't she have operationally defined that term? If she were to do so, then one of her clarifying terms might in turn require an objective definition, which also would probably necessitate another interpretive definition.

Her futile attempt to define every phenomenon unambiguously is not unique. At some point along a chain of definitions, every researcher must accept the subjectivity that occurs when communicating words and depending on primitive terms. The skilled researcher includes enough operational definitions to make identification reasonably objective, reliable and reproducible without making the descriptive definitions excessively detailed. They are overdone when decision-making becomes exceedingly tedious and/or classes are measured with more precision than necessary for solving the research problem.

Another way of gaining objectivity and consistency in field decisions (besides thoroughly describing the classes) is by making visual comparisons between each individual and the class norms and boundaries. This is essentially the same procedure as described for field identification (Chapter 3), except comparisons are not limited to only one class. For a multiple-class scheme, each individual needs to be compared with several class norms, with ranked classes requiring fewer comparisons than unranked ones. When the comparisons cannot be done directly, for example, by simultaneously looking at a model red bead and the bead to be classified, comparisons can be made with replicas or representations of the standard.

This technique is illustrated by the procedure used by Ms. S. when she classified residences in Choros. She decided to standardize field decisions by using photographs of residences that represented class norms and transitional types. During a trial run she tested the efficiency of her coding sheet (Table 5.1) and looked for residences that appeared to be a norm for each of the seven classes. She then photographed the residence that best approximated the ideal type in each class. At the time they did the actual field classifying, Ms. S. and her two friends each carried a set of these photographs.

Ms. S. asked each of her friends to classify approximately a third of the houses, but she also designated some areas of overlap for "common" evaluations. The houses in these "common" areas were judged independently by all three fieldworkers. After making their decisions and marking their coding sheets, Ms. S. conferred with the other two workers and compared the three classification decisions for each residence. If the three observers agreed, Ms. S. finalized the overall classification (type A through G, Figure 5.1) by marking it on her master map. If they disagreed, they returned together to the residence in dispute and resolved the discrepancies. Usually the differences in judgment occurred because the individual residence possessed characteristics that made it transitional between two classes. In such cases Ms. S. then took an instant-print picture of the house and marked its location on her working field map. Whenever subsequent questions arose concerning the same class boundary, this newer supplementary picture, as well as the house itself, served as a standard for making comparative judgments.

Table 5.1   Coding Sheet for Residences in Choros

Residence: A building used as a single-family home including . . .	I. D. Number _____
	Location _____
	(Street Address)
	Time of Observation _____
	Observer _____

1. Size of House:			Check one.
Large	(. . .	)	L _____
Medium	(. . .	)	M _____
Small	(. . .	)	S _____
2. Quality of siding:			Check one.
Expensive	(. . .	)	E _____
Standard	(. . .	)	D _____
3. Additional features:			Check one.
Luxurious	(. . .	)	X _____
Basic	(. . .	)	B _____

. . . = Not specified in this illustration.

Ms. S. attempted to increase objectivity in classifying inanimate objects in the field by carrying pictures of the class norms and boundary types with her so visual comparisons could be made between two observable objects. Even though one of these objects was only a photographic image, it remained constant and so provided more consistency for making comparisons than would have been the case if her standard were only a verbal description of a concept. Also, because the distance was small and the residences did not change during the time that data were collected, Ms. S. had the option of returning to the site of the standard residence if the photographs were inadequate. Returning to view the standard individuals, however, has the disadvantage that the two residences being compared can not be seen simultaneously. Furthermore, this procedure of returning to the site of the standard can not be used if the phenomenon and/or its position change rapidly.

In general, comparisons between particular individuals with standard ones may be accomplished by one or more of the following ways: (1) carrying a description of the standard to the site of the individual; (2) carrying the standard itself to the site of the individual; (3) carrying the individual itself to the site of the standard; (4) carrying a replica such as miniature, picture, sketch, or diagram of the standard to the site of the individual; or (5) viewing the standard and individual successively at different locations but within a short period of time.

Considerations in Classifying Field Phenomena

Classifying land use and other phenomena in the field is a form of data collection that is commonly used by geographers.[5] The popularity of this method is related partly to the apparent ease with which data can be acquired. However, collecting data by visually observing inanimate phenomena for the purpose of classifying individuals should be evaluated carefully in terms of possible disadvantages. Several factors that might be considered are tabled later in this chapter, but four that should receive special consideration involve (1) the reliability and reproducibility, (2) the detection of one's own errors, (3) the costs of time and travel, and (4) the potential hindrances.

The fact that almost anyone can easily observe and classify phenomena tends to lull the amateur researcher into a false sense of accomplishment. However, unless there is a consistency in measuring individuals so that nearly identical individuals are classified the same, the reliability of data may be too low for valid conclusions. Likewise, if several researchers use the same classification scheme in the same or different areas, classificatory criteria must be interpreted in an equivalent manner so the results will be comparable. If the original classification is not reproducible at a later time or different place with a high degree of reliability, then the original study loses much of its research value. Even utilizing the aids that help improve consistency and reproducibility does not guarantee that decisions made in the field will always be reliable, a fact that must be taken into consideration when judging this field technique.

Another factor closely related to consistency is the likelihood of not catching one's own mistakes. It is almost axiomatic that a person will make an error in collecting and recording data, so procedures designed to detect such errors are important. Unfortunately visual field classification, which involves making a one-time field decision, does not provide much opportunity for re-doing or systematically checking for errors. Once a classificatory decision has been recorded and the field observer has left the site of the individual, the data are usually accepted as accurate. It is very difficult later to re-examine the data and detect where errors may exist.

The costs in time and travel can be fairly high. Obviously costs depend on several conditions including the size of the study area, the number of individuals to be observed, and the complexity of the classification scheme. In any case, going to the site of each individual and observing it long enough to make a variety of judgments takes considerable time. These expenses should be realistically estimated when the researcher attempts to determine the relative merits of this technique over other data-gathering methods.

Several potential environmental hindrances should be assessed when evaluating this technique of visually classifying objects in the field. Collecting data at the sites of the individuals may be affected by outdoor temperatures, precipitation, physical barriers that restrict accessibility to individuals, hostility of the local populace, hostility of animals such as unleashed dogs, and the general well-being of the field observer. This is not to say that these possible disadvantages should deter the researcher from using this method of data collection, but the wise fieldworker will consider the possible effects of various factors on the collected data and attempt to minimize their distorting consequences.

Observing Human Subjects Unobtrusively

Collecting data by observing human beings unobtrusively is similar to classifying inanimate objects in many respects. The technique depends on the researcher's judgment about the variations in the attributes being measured because observations are done secretly without any information being consciously provided by the subject. Also, since the data must be obtained unobtrusively, the researcher must make measurement decisions in the field where the subject is located. Furthermore, when the individual being measured is an activity, the observation must be made at the time the event is occurring.

There are also differences between observing inanimate objects and human beings, and some of these differences cause additional difficulties in collecting data. These difficulties (see below) are severe enough that the researcher might initially believe it would be better to choose a data-gathering technique which asks the subjects to provide information about personal characteristics and/or behavior. Indeed, asking for information directly from the subject is an important technique (Chapter 6), but there are circumstances when unobtrusive observations may be more appropriate. Concealed observations can provide useful data when a researcher wants to be very certain that measurements of behavior are of actual behavior rather than self-reported actions. If subjects are not fully aware of all their own detailed actions, are unable to express and communicate their behavior (as with small children), or are unwilling to disclose their true activities or attributes, then disguised observations may be necessary.

The study undertaken by Ms. S. in Choros illustrates this kind of technique. One issue that kept recurring in discussions with Choros citizens was the safety of children who walked to the elementary school. As a result, Mrs. S. decided to collect information about the characteristics of schoolchildren pedestrians. The characteristics of primary concern were routes, especially where they crossed major vehicular thoroughfares, and the number of children who walked together in groups. Of the several possible techniques (see Chapter 9 for others), she chose to watch and record children's movement inconspicuously as they walked to and from school.

She planned to select intersections that were one block away from Highway #37 on the opposite side of the highway from the elementary school. She expected that as soon as a child who appeared to be an elementary pupil arrived at the intersection, she would begin observing the child's route to the highway and across it. She would also watch all other pupils that were walking with the selected child. She anticipated sitting in her car parked near the selected intersections and surreptitiously recording the data by mapping and by making explanatory notations.

This brief description of the project by Ms. S. is too sketchy to evaluate the sampling design and similar methodological decisions, but it may aid in discussing some of the limitations of the general technique. In addition to all the disadvantages associated with observing inanimate objects, this technique involves at least three other potential limitations: (1) the danger of subject awareness, (2) the ethical issue of collecting data, and (3) the complications from movement.

By definition, this technique depends on maintaining secrecy so the subjects do not modify their behavior. When measuring inanimate objects, the observer does not need to worry about being discovered; but when observing people, the researcher needs to plan ways of remaining unobtrusive. Observing persons, like watching birds, without their becoming aware of being observed is not always easy to do. In some circumstances it would be very difficult to obtain accurate data without the presence of the observer being realized by subjects. In fact, it is doubtful that Ms. S. will really be able to sit repeatedly in a parked car in this small city without attracting the attention of children and parents.

This leads to the second potential limitation, namely, the ethical issue of collecting data clandestinely. The goal of conducting valuable research into human behavior often conflicts with the goal of preserving personal privacy. How does the researcher weigh the benefits that are expected to accrue from the knowledge gained in a study against the costs of disruptions in the lives of others? This is partly solved when subjects agree to help voluntarily (Chapter 6) and have the opportunity to opt out of the project. Subjects cannot exercise this choice when they are unwitting participants. Is observing unobtrusively a form of spying? Is it, instead, like sitting in a park and casually watching strollers, or like mingling with and watching the crowds at carnivals? If people are in public areas, then don't they expect to be seen by other persons? There is no universally accepted answer to this issue. Some persons may believe it should not be listed as a disadvantage, while others may judge the technique totally unacceptable. It is important, therefore, that each researcher conscientiously consider the ethical issue of this technique within the framework of personal values. Even though some of the value judgments are especially evident in this technique, obviously the ethics of every technique involving humans must be evaluated.[6]

The third potential limitation is associated with the movement of persons. Difficulties in measuring and recording the spatial positions of moving phenomena are discussed in Chapter 9, but the distinction that is noteworthy here concerns the nature of human movements. Observations of human subjects often pertain to activities (events) rather than to more permanent attributes, as is usually the case with inanimate objects. This lack of permanence of the individual (event) combined with movement over an area make the reliability checks that are advocated for measuring inanimate objects almost impossible. For example, if Ms. S. tries this technique but then later senses that she may have made a mistake in mapping the route of student #6 on the morning of October 2nd, there is no way she can re-observe that activity. Thus, the use of class norms and boundary types (e.g., as Ms. S. used when classifying residences) is virtually impossible for many individuals in motion.

Partly because of these several limitations and partly because of the kind of phenomena commonly studied by geographers in the past, unobtrusively observing human behavior has been employed far less frequently than, say, classifying inanimate objects such as land use.[7] Nevertheless, the beginning researcher should regard this technique as one having potential merit and, thus, one to be compared and evaluated along with other techniques for gathering data about spatial behavior.

Observing Subjects as Participant Researcher

When observations are not unobtrusive and the subjects are aware of the presence of the researcher, the procedures for collecting data may differ somewhat from those used by the more isolated (unobtrusive) observer. If the researcher interacts with the subjects within their normal social environment, he/she is regarded as a *participant observer.*[8] Rather than observing characteristics and behavior entirely from a secluded viewpoint, the researcher attempts to become a part of a social group or situation to gain a more internal viewpoint. By participating with the subjects, the researcher hopes to gain understanding by actually experiencing some of the perspectives of the persons being observed. It is expected that these experiences all aid in collecting information that pertains to complex social attributes which are not so apparent to an outsider. Furthermore, by functioning as a participant, the researcher can combine data measured through visual observations with information acquired by asking subjects about various facts, including intangible attitudes, feelings, and beliefs. Thus, by expanding the types of data gathered beyond merely visual observations, the researcher may be able to learn about obscure attitudes, intense feelings, and a variety of emotions which may provide insight into relationships that will strengthen the interpretation of the data.

There are many variations in this set of data-gathering techniques. They differ, among other ways, according to the involvement of the observer and the extent to which subjects are aware of the researcher's purpose for participation. Because of considerable variation, sometimes this general method is divided into two subtypes: quasi-participant and full participant.[9]

Quasi-Participant

A quasi-participant is an observer that socially interacts with the subjects but remains an outsider. The researcher in this role is primarily an observer but one who is known and accepted by the subjects. This technique may be further subdivided according to the subjects' awareness of the researcher's objectives. One subdivision includes procedures for collecting data without revealing the purpose of involvement. The technique remembles that of unobtrusive observation—except the observer is not physically hidden. The other subdivision occurs when the subjects are cognizant that the researcher is collecting information. The technique then resembles unstructured interviewing (Chapter 6).

The first subdivision, in which the research goals are unknown to the persons being observed, is more commonly associated with the term quasi-participant (or marginal participant). It is illustrated by the strategy Ms. S. adopted after she abandoned the unobtrusive one. First she explained her project to Dr. A. and the head of the P. T. A. They both agreed that she could work with parents who supervise the highway school crossings. When she was ready to collect data, she joined the parents who were assigned to the designated school crossing of Highway #10 located directly south of the school building. She performed as other adults who assisted children at the school crossings during the morning and afternoon hours, but she also gathered data by recording selected observations on note cards. Some of the children sensed that she was not "a mother", but they generally accepted her presence as "belonging" and acted as they did around other adults.

The fact that Ms. S. is not a child made it impossible for her to integrate completely with the pupils. This situation resembles other circumstances where a researcher is unable to become an indistinguishable member of a population due to barriers of age, sex, language, or cultural background. In spite of differences and the subjects' vague awareness of being observed, Ms. S. attempted to minimize her influence on the normal spatial behavior of those being observed. She attempted to function like other persons who interacted with the children so that her role as an observer did not alter the subjects' usual behavioral patterns. Her success at being accepted into the subjects' social environment without unduly affecting their normal spatial activities was probably more dependent on her personal qualities than on specific strategies used.

Whenever a researcher attempts to function as a quasi-participant without revealing the true and major reason for being in the group, the recording of data must necessarily be very inconspicuous. Surreptitiously jotting brief notes on cards or speaking quickly into a tape recorder may serve as temporary data storage until there is an opportunity to expand, organize, and clarify these abbreviated field records. In some situations the researcher may be unable to use memory aids and the recording of data will have to depend on recall after the group activities are completed.

The second subdivision pertains to members of the target population when they know they are being observed by a researcher (a recognized outsider). Although they may not know the researcher's exact goal, their awareness of being watched will probably alter their behavior. Such modifications, called the Hawthorne effect, were illustrated by a study of workers in a factory (specifically, an electric company's Hawthorne Plant in Chicago in the 1930s).[10] The researchers noted that production increased when lights in the factory were increased and also when they were decreased. Similarly workers reacted in the same way to varying but contrasting kinds of rest periods. The investigators concluded that the increased production resulted from the workers' awareness of being watched rather than from changes in lighting or rest periods.

Full Participant

A full participant attempts to gain understanding about human phenomena by becoming a functioning member of the studied population. The role of the researcher usually is unknown, although in some studies a few subjects may possess a vague awareness of the observer's intentions. By fully participating as a member of the group, the researcher seeks to combine visible activities with intangible meanings, feelings, and attitudes that other members experience.

Observation by a full participant is illustrated by a project undertaken by Mr. C. He felt that more families favored the present site for the Choros elementary schoolbuilding than those who actually expressed their views at a recent public meeting. He believed that many persons who favored the present site did not attend the meeting because they were working, felt diffident about speaking in public, and/or believed their opinions wouldn't make much difference. He decided to "feel the public pulse" by learning about the opinions of as many citizens as possible within a month's period. Because of his own vigorous advocacy of one point of view, he thought some persons might be reluctant to express their true feelings to him. Therefore, he persuaded a friend and his wife (Mr. and Mrs. G.) to do the asking. They were to inject the schoolhouse issue into conversations whenever possible at various gatherings in hopes that others would reveal their opinions. In one respect the illustration varies from a typical participant observer because Mr. and Mrs. G., who were already participants, became researchers. This contrasts with the usual process of a researcher becoming a participant. Nevertheless, the effects are similar. Mr. and Mrs. G. interacted with the persons who possessed the data they wanted. They extracted from these contacts certain information that was to be used to meet research objectives.

Some Limitations

Some of the conditions that reduce the utility of other observational methods also apply when the observer becomes a participant. For example, the time element is a deterrent because the cost of obtaining data as a participant observer is especially high in terms of time. This includes the time spent in preparing to learn the role of the subjects, arranging for entry into the population, and gaining skill at eliciting the desired data. Mr. and Mrs. G. had it easy because they were already members of groups to be studied, but many researchers face formidable obstacles in gaining the skill required to imitate members of an unfamiliar group. For some anthropological studies (i.e., not necessarily geographic ones) this method has required years of time.[11]

Another limitation to participant observations is the possible loss of reliability because of the subjectivity of measurements. In effect, the researcher abandons the role of an objective observer of a phenomenon and becomes a part of that being measured. When the goal is to join (infiltrate) the target population to such a degree that the new participant's role as a researcher is unknown to the subjects, the researcher-participant must outwardly adopt many of the characteristics of the members of that population. Often various inward adoptions occur also, and the researcher must be cognizant of them. The participating researcher must be aware of having to make judgmental decisions or measurements about a phenomenon that includes self-involvement. The lack of objectivity (or reproducibility by another scholar) may be especially severe if the researcher becomes emotionally involved with the group.

Also, to become a part of the target population means interacting with individuals. This in turn means influencing their behavior and attitudes. How can Mr. and Mrs. G. get their acquaintances engaged in revealing conversations without also voicing some ideas, which may then alter the views of the subjects? To what extent do secret agents who infiltrate political organizations for the purpose of gaining information actually promote activities that are to be observed? How can the researcher separate the characteristics of the observed individuals, particularly individual events, that are "normal" ones from those affected by the very act of studying the individuals? It is almost impossible to determine the degree to which the phenomenon being measured is altered by the influence of the researcher. Even though the effects cannot be known, researchers should be aware of the dangers of affecting the observed phenomenon, and they should attempt to minimize such potential influences.

Comments on Experiential Fieldwork

Additional evaluations of the various techniques grouped under the title of participant observations involve some basic research assumptions, particularly ones associated with objectivity in studying human phenomena. In Chapter 1 empirical science is introduced as the philosophy that undergirds geographic research and justifies the study of field techniques. Throughout the book the stress is on procedures that achieve objectivity in measuring phenomena. It is important, however, to emphasize both the impossibility of achieving perfect objectivity and the limitations of empirical modes of acquiring knowledge. Although the issues concerning empiricism versus other philosophical approaches concern all aspects of knowing, and hence pertain to all techniques used in studying humans, reasons for challenging the methodology of empirical science are most obvious when one evaluates the merits of participant observations.

As described above, the researcher who becomes a participant observer seeks to gain information about other persons, not only by observing visible and audible stimuli but also through an awareness of their feelings and other intangible attributes. To achieve this goal, the researcher must learn how members of the target population organize, categorize, and symbolize the infinite variations in their environment, and he/she needs to develop an empathic awareness of the emotional worlds of the subjects. These objectives cannot be accomplished without meaningful social interaction, which requires the researcher to participate in a form of fieldwork that is *experiential*.

Some limitations to participant observations are given in the preceding section, but here a few fundamental issues are re-examined or expanded to include aspects that affect, to varying degrees, all field techniques. For one, the loss of objectivity is mentioned above as a weakness associated with observing human phenomena as a participant, but this association occurs primarily because researcher subjectivity is more obvious than in many other techniques. No person commences a study in a *tabula rasa* condition because everyone is a unique product of a lifetime of personal experiences. Whether accepting the primitive terms in a classification guide or deciding on the responses choices in a questionnaire, the researcher is injecting a set of personalized concepts and perceptions into the data-gathering process. Therefore, to imply that the empirical scientist can study human phenomena as an objective "outsider" is to misunderstand the nature of the inescapable subjective bias that infiltrates all studies.

Another limitation previously discussed is the effect the researcher may have on the subjects. When the researcher is a participant observer, it is fairly easy to see how the respondent's behavior and conversations might be altered by the actions and remarks of the fieldworker. Changes also occur even during brief interviews and similarly structured communications. Two persons who interact are never the same as they were before they encountered, evaluated, and reacted to each other. The respondent who answers the last question in a formal interview is not exactly the same person who answered the interviewer's first question a few minutes earlier.

In addition to the immediate reactions, some research procedures produce long-range changes, which in turn involve ethical issues. How many people have been disillusioned because they expected to gain benefits by cooperating with research projects, but were later disregarded by a researcher who was mainly eager to collect data for personal objectives? Because the potential for influencing subjects is greatest when the social interaction is most intense and prolonged, consideration of long-term effects is most pronounced with experiential field techniques.

The changes resulting from social interaction with members of a target population also affect the researcher. Some of these are the cumulative experiences of learning more about the phenomenon being studied. These are most apparent when a researcher can review personal notes that have been logged in a diary which the researcher has maintained beyond the stage of field reconnaissance. Changes occur also because social research itself produces stress. There is the anxiety of interacting with strangers. For many amateurs it is difficult to appreciate fully the fact that the subjects hold the dominant position, a position resulting because they possess the knowledge desired by the researchers. This situation often conflicts with an image of themselves as investigators in a dominant capacity. This reversal of the role expected by the researcher may create frustration and internal conflicts. Thus, to engage in research of human phenomena is to experience changes. The researcher no longer views the world the same as before the latest experience. Indeed, it is illusory to believe that one can remain an unchangeable and unchanged "objective" observer while interacting with other humans.[12]

Because of the tremendous amount of individualized adjustment and adaptation required from researchers who study human phenomena experientially, some scholars have even questioned whether skills for conducting interactive types of research can be transmitted by textbooks and field courses. Obviously the author of this text believes general guidelines can be conveyed through the written language. It is hoped that readers are, indeed, learning about ways scholars have attempted to improve the quality of their data. Nevertheless, the aspiring researcher should be aware that the personal ingenuity and individualized modifications that often supplement basic field techniques are usually the result of direct experience. Furthermore, these individualized skills, which are most essential for collecting data by participant observations and similar experiential methods, are beneficial to all data-gathering techniques.

Observing Human Subjects within Controlled Conditions

Usually a geographer strives to observe phenomena, including human beings, in their normal settings which have not been changed by the researcher's presence. An investigator may, however, purposely alter the subjects' environments because the extreme complexity of social behavior often frustrates attempts to isolate relationships among a few specified variables. For this reason, it may be advantageous for a researcher to control and manipulate selected variables for the purpose of observing the resulting changes in human behavior.

This section groups together those techniques that involve an intentional modification of the subjects' environment for the purpose of controlling some of the conditions within which persons make locational decisions. The extent of control may range from minor alterations in the field to highly regimented conditions in a laboratory. The contrasts are great enough that they justify discussing each of these two types separately.

Controlled Field Conditions

Observing human subjects under controlled field conditions is very similar to the unobtrusive techniques discussed above, except the researcher attempts to discover relationships by altering selected phenomena. The contextual situation in which geographic research is conducted makes it extremely difficult to isolate the interrelationships among variables. One way of detecting spatial relationships is by varying one phenomenon and then observing the consequent changes in other variables. The goal is to intrude variably into an existing situation in a manner that encourages subjects to continue behaving as usual, except for adjustments to the modified spatial context.

Dr. A. was worried about the traffic near the site of the present elementary school building. She was eager to obtain data about current traffic flow and about possible changes that could be achieved without closing major thoroughfares. She explained her concerns to the proper city officials and asked for their help. They authorized Mr. D., the director of safety affairs, to obtain data about the possible effects of changing traffic patterns near the school.

Mr. D. worked out a scheme for prohibiting vehicular traffic during a prescribed period on specified blocks. He then physically closed various streets as he tried different patterns of restrictions. Mr. D. measured the number of vehicles and their speed at a set of twenty-two street positions. For data on the number of cars, he used automatic traffic counters. For the speed, he utilized radar equipment. In both cases he depended on the reliability of the physical instruments and his training to operate them correctly. For the locational data he estimated the recording positions on large-scale maps that showed the relevant streets. Since the data referred to traffic along an entire block (with minor exceptions), the precise positions of the data-gathering sites were not critical.

Some citizens couldn't understand why "the city" closed some streets—especially when no one seemed to be repairing potholes—but they obeyed the directions and drove on other streets. There was no evidence that anyone deliberately drove back and forth on a particular street to affect the automatic counters. Therefore, the technique appeared to be an accurate indicator of traffic patterns that might be expected if specific streets near the school building were to be closed permanently. Dr. A. realized that predicting the future is always risky because the sampled population would not necessarily be identical to the future population. Even so, the results provided a good indication of behavior when specific conditions were altered.

The way Dr. A. later used the traffic data resembles the technique of simulation because she tried to simulate or copy the street pattern that would exist if certain blocks were permanently closed. She then used these data as a base for predicting future traffic flows and, by implication, their possible effects on pupils. Another researcher might have used similar methods to obtain traffic data, but for the purpose of establishing a relationship between human characteristics and the way the people use city streets. Irrespective of the research emphasis, the method of unobtrusively observing human subjects under controlled field conditions is a useful data-gathering device.[13]

In spite of the attractiveness of this technique as a way to control or account for selected variables, geographers seldom utilize this method because of the complexity of spatial interrelationships. Locations of phenomena on the earth's surface result from so many natural and human forces that it is extremely difficult for a single person to control all the individuals belonging to the relevant phenomena. Not too many persons are in a position of power, even over a small area like Choros and over a minor alteration such as street accessibility, that allows them to change the locations of an entire set of individuals.

Controlled Laboratory Conditions

Social scientists in several disciplines sometimes attempt to study behavior by asking subjects to perform specific tasks while being observed in a small room or laboratory setting. By isolating volunteers from more complex social environments, researchers seek to emulate the controlled laboratory conditions of other scientists who study natural phenomena.

This technique of observing persons under controlled laboratory conditions contrasts in several ways with the procedures used in the field. One of the primary contrasts is in the awareness of subjects. It is virtually impossible to bring persons into a laboratory without their knowing about their role as data generators. It may be that the volunteers are not told which specific characteristics will be observed; but nonetheless, they are cognizant of the artificiality of the setting. Their normal behavior is, therefore, affected. Thus, the researcher is confronted with possibilities that the act of measuring individuals may itself influence the individuals in a way that produces atypical data.

Another contrast is the removal of subjects from the context of their outdoor setting. In fact, the domain of geography is so inextricably related to areas outside of buildings that spatial phenomena occurring within rooms are usually omitted from geographic studies. This is not to imply that studies about the use of interior (room) space cannot contribute to understanding the use of earth-bound environments, but the magnitude and nature of the contribution is difficult to assess.[14] In fact, for geographers the greatest drawback to this technique is the uncertainty about the applicability of conclusions based on spatial behavior inside laboratory areas to the locational actions of persons outside the confines of buildings.

Comparing Control and Subject Awareness

Methods for observing phenomena visually have been grouped in this chapter into four (six, counting subdivisions) classes for presentation. These were organized according to increasing amounts of research control, commencing with situations where the observer watched inconspicuously and ending with the observer controlling many of the pertinent variables. With increasing control the solution to the research problem might be expected to become easier because certain variables can be held constant or isolated for measurement and hence aid in establishing relationships.

Increased control, however, may be gained by compromising other objectives, one of which is to obtain information about the subjects' characteristics that are true or unaltered. Subjects may change their characteristics and behavior if they become aware of being observed. For example, television reporters and camera-persons attempt to observe and record information without involvement or affecting the action of being reported. It appears, however, that some subjects, when they become aware of being observed by a television camera, greatly alter their behavior.

The benefits of researcher control do not correspond exactly with the costs of awareness and altered data. The compromises may occur in various combinations. The relationships can be diagrammed—using a very imprecise measurement—by comparing these six methods on scales of control and awareness (fig. 5.2). Although the relationships are only suggestive, the diagram does display the comparative advantages of the technique that unobtrusively controls selected variables in the field (CF). It allows considerable control of variables but without much sacrifice in subject awareness.

Ways of Recording Data

This chapter concerns techniques requiring the observer to measure data in the field with the only record of those measurement decisions being reports completed while observing phenomena. For many phenomena, especially those in motion, the observation is made only once and can never be rechecked. For others it may be difficult and costly to return and recollect data. Because there may be no second chance for observing and measuring the phenomena, the manner in which the data are recorded is very inportant.

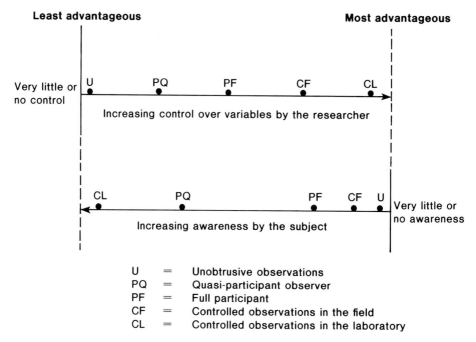

Figure 5.2. Comparisons of observational techniques according to control and awareness

General Characteristics

The overall purpose of field records is to communicate to oneself a series of spur of the moment decisions made at one time but to be interpreted at a later time. The communication should be unequivocal, as complete as possible, and convenient. It should minimize the filtering that inevitably occurs. Unequivocal records should be clear, legible (or audible), and without extraneous markings (noise). There should be no confusion about the association of facts with their respective places.

Data sheets should be designed to insure completeness. Data gaps, which may result because a few individuals cannot be measured, are expected in some situations. Incompletes should not occur, however, because of oversight errors. A well-designed format for recording data will make gaps appear obvious and will aid the researcher in catching any errors of omission.

Recording forms should be convenient to use in the field. This rule applies especially to adverse weather conditions for outdoor observations, and to all situations where concealment is critical. Recording formats should also be conveniently arranged for rapid use. Considerable time is required for collecting data by most methods of visual observations. To minimize recording time, codes, symbols, check-marks, or similar shorthand should be used.

Data should be arranged conveniently on record sheets for storing, retrieval, and transfer. Because the lapse in time from collection to utilization is frequently long enough to forget specific facts, stored data must be labeled completely by title, location, and date so it can be retrieved accurately. The positioning of data on record sheets can affect the rate and accuracy of data transfer, whether by key-punching or by hand tabulations. In general, the researcher should design a format that is convenient for organizing and working with data in all stages of the project.

Maps

Maps, a universal tool in geography, serve as a valuable form for recording data. Base maps are very versatile in the kinds of information that can be recorded. Maps are abstractions in themselves; therefore, additional coded and symbolized field notes are appropriate forms for compact data. For recording the facts of location associated with each fact, maps are very efficient. It is usually much easier and faster to place a symbol at the correct position on a map than to describe such a location verbally. Also, the completed map may serve not only as a form for recording data, but it can be a first step toward spatial analysis.

Base maps do have limitations when used as data sheets. One disadvantage is the limited space available for multiple facts per areal unit. Unless the map is extremely large scale, there is seldom room to place anything more than a few symbols at each datum site. Another disadvantage is that, although the arrangement of mapped data is suitable for immediate areal analysis, it can be cumbersome to transfer the data to another form. If, for example, the results were to be tabulated in a column during the analytical stage, the researcher would have to develop a systematic procedure for finding all the data scattered areally across the map. A third consideration, which unfortunately must be placed with these potential disadvantages, is the difficulty many persons have in reading and constructing a map accurately. If researchers realize they are likely to make errors in locating data correctly on a base map, they should use another form for recording the data and their locations.

Sketches and Photographs

Less abstract forms than a map for recording data are series of sketches or photographs. These forms attempt to reproduce observed characteristics, for later re-examination at a study desk. Sketches are partial abstractions because the observer omits many phenomena, yet the recorded information usually appears more realistic than that conveyed by map symbols. Photographs are even less abstract so their form permits rechecking many facts without returning to the field. Pictures may not always be worth a thousand words, but they certainly contain a lot of information in a compact form.[15]

A limitation of sketches or pictures is their restriction to a single viewpoint. This perspective prevents recording many characteristics, including most two-dimensional measurements. For example, the photographs Ms. S. took of selected houses in Choros were helpful in comparing quality of siding, but they were incomplete records for measuring houses by size. If she had taken a series of pictures or used videotapes that showed several perspectives of each individual observed in the field she could have overcome this limitation. For individuals in motion, recording with motion photography (movies or videotapes) is a practical medium.

Another limitation is the cost in time for sketching and the cost in equipment for photographing each individual (e.g., each residence in a section of Choros). Furthermore, sketching may not be feasible for researchers who have little skill in recording visual phenomena in this manner.

Like all unmapped data forms, pictures require the recording of additional data about location. The geographic element is included easily when recording data on maps because field locations are concurrently measured when data about other relevant characteristics are placed on the map in the correct position. The locational information, therefore, is an integral part of the recorded data. In contrast, pictures and other non-mapped data forms require locational facts to be tied unequivocally to each fact. When using pictures and all other non-map forms, therefore, a data-recording procedure should be established in such a way that locational facts are never forgotten and are accurately connected to the other data.

Codes and Notes

Abbreviations and codes are to written records as symbols are to maps. They are essential for recording data quickly and for conveying meaning in a precise but concise manner. One of the most common forms for recording data is a sheet on which the observer codes the information by adding or checking symbols (e.g., Table 5.1). Any set of symbols that can be memorized easily and/or consulted quickly in the field may be adequate for recording data. When devising a coding scheme, however, their subsequent use should also be considered. For example, if the researcher planned to transfer data from field coding sheets for computer analysis, conversion errors might occur if the symbols Ψ, ∇, and $\exists$ were used in the field rather than symbols compatible to the computer.

The electronic equipment of today allows researchers to code data directly for computer analysis without involving the transfer process from coding sheet to computer cards or terminals. This is because the output of an electronic coder is in a form compatible with computer language. Combined with this advantage is the ease by which observers can watch the target phenomena and record their decisions while punching the numbers and symbols on the hand coder.

Coding is so common that it is frequently combined with other forms such as maps. An example of this is the system developed for recording data in an area under the jurisdiction of the TVA in the 1930s. The system used a "fractional" code to be placed on base maps.[16] The top portion was a sequence of digits that symbolized classes of land use, crops, field size, amount of idle land, and farmstead quality. The bottom portion of the "fraction" indicated the classified characteristics of the natural environment (e.g., slope, drainage, soil depth) by an ordered set of digits. The field coding for a particular minimum areal unit might appear on the base map as 3.1.4.2.2 above the "fractional" line and 1.4.1.1 below the line.

Field notes function like codes because they are a form of communication written by the researcher in the field. The two forms differ in their relative merits. Since notes consist of more complete words and phrases than code symbols, they take longer and more space to record.[17] This disadvantage may be more than balanced by the benefits associated with field notes. One of these benefits involves a continuation of the reconnaissance when hypotheses and relationships are still unclear to the researcher. This is especially true for participant observations where the researcher has to become a member of the target population before being able to know the variables which are most meaningful and hence need to be observed more carefully. Frequently participant researchers record their observations in the form of field notes.

In other circumstances, field notes are beneficial when they supplement other data forms. Keeping a field log that clarifies symbols, explains gaps, elucidates exceptions, and interprets the record is a very valuable technique. An explanatory comment written to oneself while observing an individual in the field may prevent uncertainty about a specific marking later when memory has failed. Researchers who, while analyzing their data, encounter difficulty interpreting some symbols and must return to the field to clarify their meaning may wish that they had originally spent a little extra time in the field to write explanatory field notes.

Field "notes" normally refer to written comments, but oral remarks recorded on tape can serve the same purpose. A running conversation with oneself about how particular measurements were made would be very helpful later if it has been preserved on tapes which can be replayed when the data forms are being analyzed.

Evaluation and Applications

Evaluation

Some advantages and disadvantages accompany the presentation of the various observational methods in this chapter. These evaluative comments deal with primary factors that distinguish these few methods from each other in terms of their suitability for gathering data. However, many other factors should also be considered by the researcher who wishes to judge the comparative strengths and weaknesses among several possible techniques. Furthermore, these factors usually apply to most field techniques, not just to those involving visual observations. It is appropriate, therefore, to consider a check list that summarizes the many questions a researcher should answer when choosing a technique for data collection (Table 5.2). The purpose of the list is to remind a researcher of the many ingredients that may affect a particular technique and, thus, aid or handicap the collection of accurate and meaningful data.

Although the questions in Table 5.2 are phrased so an affirmative response implies an advantage, a negative reply should not be regarded as a deterrent to its acceptance. Answering "no" to a specific question may indicate only a partial limitation or disadvantage that must be endured to gain the expected benefits. Often a meaningful reply for a particular technique is neither an absolute yes or no, but rather an answer which indicates a response that is more or less affirmative than can be given for an alternative technique. The primary function of the guide, therefore, is to remind prospective researchers of several criteria to be considered when evaluating various field techniques.

The check list may serve as a guide at several stages of research planning. It may help during the preparatory stage when a researcher is first considering ways of measuring phenomena and collecting data. At this stage only generalized comparisons among several techniques are possible because detailed procedures depend on operational definitions, sampling design, and similar decisions. Later, after specific procedures for collecting data have been planned and tried during a pilot project, the list can be re-examined and the technique re-evaluated in terms of its detailed procedures.

Table 5.2 A Guide for Evaluating Field Techniques

Criterion	Question
A. Measurement Characteristics	
1. Validity	Am I reasonably certain that what I plan to measure corresponds to the research phenomenon?
2. Reliability and reproducibility	Will my measurements be consistent? Would other persons get the same results (subject to sampling variability) if they followed my procedures?
3. Standardization	Are the variables measured in a manner that has been standardized by other researchers and hence are comparable?
4. Precision	Will my measurements be as precise as required to solve the research problem?
5. Detection of errors	Is it possible for me to catch my own omissions and errors committed while collecting the data?
6. Locational facts	Can each fact be easily associated with corresponding data about its location?
B. Control	
1. Control of variables	Does this technique allow me to alter or hold constant any or all of the variables?
2. Control of sample	Can I control the sampling procedure to guarantee randomness?
C. Researcher-Subjects Relationship	
1. Awareness	Can I prevent human subjects from being aware of my collecting data?
2. Cooperation	Is the technique independent of any required cooperation from the human subjects?
3. Effects	Can I avoid affecting the human subjects in a way that would, in turn, alter the data?
4. Ethics	Does the procedure respect the human rights of the subjects?
D. Scale	
1. Size of study area	Can I collect data from a large area during a short period of time?
2. Locational precision	Can I achieve adequate locational precision in the geographic positions and, if continuous phenomena, in the areal units?
3. Sample size	Can I collect data for as many individuals as necessary to achieve specified sampling objectives?

Table 5.2—*Continued*

E. Cost
  1. Time

Can the data be collected within a reasonably short period of time?

  2. Travel

Can the data be collected without having to spend too much for travel?

  3. Equipment

Can the data be collected without any special equipment? If not, is the equipment inexpensive and readily accessible?

  4. Assistants

Can I collect the data without having to depend on paid assistants?

  5. Training

Is the collecting procedure simple? Can I and/or assistants make accurate measurements without special training?

  6. Hindrances

Is the technique free of natural and human conditions that might cause delays, disruptions, irritations, and other hardships?

F. Recording
  1. Form

Is there a recording form that is appropriate for the data?

  2. Format

Can the reporting format be organized for easy, consistent, and accurate use in recording and later interpreting and transferring data?

Applications

Most of the publications listed in the Notes refer to the application of one or more of these observational techniques, so additional information about them can be gained by reading the reports. For more personal familiarity with some of the techniques, you might do one or more of the following projects.

1. If you designed a classification scheme for measuring visual attractiveness (as suggested at the end of Chapter 3), try using this scheme in the field to classify individual areas according to their visual attractiveness. If you can persuade a friend to use your scheme for classifying the same areas independently, you can compare your friend's results with your own. The comparison should indicate to you where your classification scheme and its field utilization need improvement.

2. Collect data on some spatial activity such as the movement of persons in a student union, a cafeteria, a bar, or other public area. Try collecting data twice: once unobtrusively and a second time when your acts of observing and taking notes are conspicuous. List the difficulties you experienced with one technique that did not occur when you used the other technique.

Notes

1. G. J. McCall and J. L. Simmons, eds., *Issues in Participant Observations: A Text and Reader* (Reading, Mass.: Addison-Wesley, 1969), p. 61.
2. R. L. Gordon, *Interviewing: Strategy, Techniques, and Tactics,* rev. ed. (Homewood, Ill.: Dorsey Press, 1975), p. 33.
3. This is written to include persons with visual disabilities too because they use alternative senses to gain much information about phenomena which is visible to others.

4. C. L. Lastrucci, *The Scientific Approach: Basic Principles of the Scientific Method* (Cambridge, Mass.: Schenkman, 1967), p. 162.

5. C. Board, "Field Work in Geography, with Particular Emphasis on the Role of Land-Use Survey," in R. J. Chorley and P. Haggett, eds., *Frontiers in Geographical Teaching* (London: Methuen, 1965), pp. 187–214, especially pp. 195–204; J. Lounsbury, L. Sommers, and E. Fernald, *Land Use: A Spatial Approach* (Dubuque, Iowa: Kendall/Hunt Publishing Company, 1981); A. J. Hunt, "Land-Use Survey as a Training Project," *Geography*, 38 (1953), pp. 277–286; G. D. Hudson, "The Unit Area Method of Land Classification," *Annals, Association of American Geographers*, 26 (1936), pp. 99–112; R. E. Murphy, J. E. Vance, Jr., and B. J. Epstein, "Internal Structures of the CBD," *Economic Geography* 31 (1955), pp. 21–46; R. Murphy, "Urban Land-Use Maps and Patterns," in *The American City: An Urban Geography* (New York: McGraw-Hill, 1966), pp. 186–206; *Rural Land Classification Program of Puerto Rico* (Evanston: Northwestern University Studies in Geography, 1952); L. D. Stamp, *The Land of Britain: Its Use and Misuse* (London: Longmans Group, 1948).

6. K. T. Erikson, "A Comment on Disguised Observation in Sociology," *Social Problems*, 14 (1967), pp. 366–373; R. L. Gorden, *op. cit.*; L. L. Horowitz, ed., *The Rise and Fall of Project Camelot: Studies in the Relationship Between Social Science and Practical Politics* (Cambridge, Mass.: M. I. T. Press, 1967); J. Katz, *Experimentation with Human Beings* (New York: Russell Sage Foundation, 1972); H. C. Kelman, *A Time to Speak: On Human Values and Social Research* (San Francisco: Jossey-Bass, 1968); J. A. Roth, "Comments on 'Secret Observation'," *Social Problems*, 9 (1962), pp. 283–284; G. Sjoberg, ed., *Ethics, Politics, and Social Research* (Cambridge, Mass.: Schenkman, 1967).

7. Some publications on this topic, not necessarily by geographers, include: D. Garbrecht, "Pedestrian Paths through a Uniform Environment," *Town Planning Review*, 42 (1971), pp. 71–84; R. G. Golledge and G. Zannaras, "Cognitive Approaches to the Analysis of Human Spatial Behavior," in William H. Ittelson, ed., *Environment and Cognition* (New York: Seminar Press, 1973), pp. 59–94; E. T. Hall, *The Hidden Dimension* (Garden City, N.Y.: Doubleday, 1966); S. John Hutt and C. Hutt, *Direct Observation and Measurement of Behavior* (Springfield, Ill.: Charles C. Thomas, 1970); R. Sommer, *Personal Space: The Behavioral Basis of Design* (Englewood Cliffs, N.J.: Prentice-Hall, 1969); and E. J. Webb, D. T. Campbell, R. D. Schwartz, and L. Sechrest, *Unobtrusive Measures: Non-reactive Research in the Social Sciences* (Chicago: Rand McNally, 1966).

8. H. S. Becker and B. Geer, "Participant Observation and Interviewing: A Comparison," *Human Organization*, 16 (1957), pp. 28–32; H. S. Becker, "Problems of Inference and Proof in Participant Observation," *American Sociological Review*, 23 (1958), pp. 652–660; J. H. Burnett, "Ceremony, Rites, and Economy in the Student System of an American High School," *Human Organization*, 28 (1969), pp. 1–10; R. W. Haynes and A. F. Zander, "Observation of Group Behavior," in L. Festinger and D. Katz, et al, eds., *Research Methods in the Behavioral Sciences* (New York: Holt, Rinehart and Winston, 1966), pp. 381–417; J. M. Johnson, *Doing Field Research* (New York: Free Press, 1975); B. W. Junker, *Field Work: An Introduction to the Social Sciences* (Chicago: University of Chicago Press, 1960); G. J. McCall and J. L. Simmons, eds., *op. cit.*; and L. Schatzman and A. L. Strauss, *Field Research: Strategies for a Natural Sociology* (Englewood Cliffs, N.J.: Prentice-Hall, 1973); J. Zeizel, *Inquiry by Design: Tools for Environmental-Behavior Research* (Monterey, Calif.: Brooks/Cole, 1981).

9. Some social scientists have classified the variations in other ways, for example, into one of four types: (1) complete participant, (2) participant as observer, (3) observer as participant, and (4) complete observer; see R. L. Gold, "Roles in Sociological Field Obmervations," *Social Forces*, 36 (1958), pp. 217–223.

10. F. J. Roethlisberger and W. J. Dixo, *Management and the Worker* (Cambridge, Mass.: Harvard University Press, 1939).

11. A. B. Hollingshead, *Elmtown's Youth* (New York: Wiley, 1949); O. Lewis, *Life in a Mexican Village: Tepotzlan Revisited* (Urbana: University of Illinois Press, 1951); F. Malinowski, *Coral Gardens and Their Magic: A Study of the Methods of Tilling the Soil and of Agricultural Rites in the Trobrian Islands* (London: Allen & Unwin, 1935); M. Mead, *The Changing Culture of an Indian Tribe* (New York: Columbia University Press, 1932); and R. Redfield and R. A. Villa, *Chan Kom: A Maya Village* (Washington, D.C.: Carnegie Institute of Washington, 1934).

12. For additional reading on the issue of "objectivity", see R. A. Georges and M. O. Jones, *People Studying People* (Berkeley: University of California Press, 1980) and G. D. Rowles, "Reflections on Experiential Field Work," in D. Ley and M. S. Samuels, *Humanistic Geography: Prospects and Problems* (Chicago: Maaroufa Press, 1978), pp. 173–193. Other sources are H. S. Becker, "Whose Side Are We On?" *Social Problems,* 14 (1967), pp. 239–247; R. Bogdan and S. J. Taylor, *Introduction to Qualitative Research Methods: A Phenomenological Approach to the Social Sciences* (New York: Wiley, 1975); S. T. Bruyn, "The New Empericists: The Participant Observer and Phenomenologist," *Sociology and Social Research,* 51 (1967), pp. 317–322; W. J. Filstead, ed., *Qualitative Methodology: Firsthand Involvement with the Social World* (Chicago: Markham, 1970); A. W. Gouldner, "Anti-Monotour: The Myth of a Value-Free Sociology," *Social Problems,* 9 (1962), pp. 199–213; J. Lofland, *Analyzing Social Settings: A Guide to Qualitative Observations and Analysis* (Belmont, CA: Wadsworth, 1971); and Note #4 in Chapter 1.

13. D. N. Chambers, *Buses and Pedestrians: An Evaluation for the Royal Borough of New Windsor* (Reading: Local Government Operational Research Unit, Royal Institute of Public Administration, 1973); A. Katz, D. Zaidel, and A. Elgrish, "An Experimental Study of Driver and Pedestrian Interaction During the Crossing Conflict," *Human Forces,* 17 (1955), pp. 514–527; P. L. McGrew, "Social and Spatial Density Effects on Spacing Behaviour in Preschool Children," *Journal of Child Psychology and Psychiatry and Allied Disciplines,* 11 (1970), pp. 197–205; and E. Sundstrom, "Crowding as a Sequential Process: Review of Research on the Effects of Population Density on Humans," in A. Baum and Y. M. Epstein, eds., *Human Response to Crowding* (New York: Wiley, 1978), pp. 31–116, especially Table B, pp. 92–101. For aspatial problems, see D. T. Campbell and J. C. Stanley, *Experimental and Quasi-Experimental Designs for Research* (Chicago: Rand McNally, 1971) and R. Plutchik, *Foundations of Experimental Research* (New York: Harper & Row, 1968).

14. A. Baum and Y. M. Epstein, eds., *Human Response to Crowding* (New York: Wiley, 1978); R. B. Bechtel, "Human Movement and Architecture," in H. Ittelson, and L. G. Rivien, eds., *Environmental Psychology: Man and His Physical Setting* (New York: Holt, Rinehart and Winston, 1970), pp. 642–645; Y. M. Epstein and R. A. Karlin, "Effects of Acute Experimental Crowding," *Journal of Applied Social Psychology,* 5 (1975), pp. 34–53; and E. Sundstrom, *op. cit.,* especially Table A, pp. 74–91.

15. J. Collier, Jr., *Visual Anthropology: Photography As a Research Method* (New York: Holt, Rinehart & Winston, 1967).

16. J. Lounsbury, L. Sommers, and E. Fernald, *op. cit.;* G. D. Hudson, *op. cit.,* and J. F. Lounsbury and F. T. Aldrich, *Introduction to Geographic Field Methods and Techniques* (Columbus, Ohio: Charles E. Merrill, 1979).

17. For an illustration of field notes, see J. Zeisel, *op. cit.,* pp. 120, 121.

Chapter 6

Collecting Data by Asking Questions

This chapter continues the discussion of techniques for obtaining data by direct observation, but the emphasis is on sensing information by listening rather than by looking. The main concern here is with gathering data that are expressed by human beings about themselves and their views on other phenomena. The division between the previous chapter, which stresses visual observations, and this chapter is a transitional one because many techniques depend on the researcher's use of both visual and auditory senses. For example, participant observers frequently learn many facts from eavesdropping and engaging in normal conversation in addition to asking purposeful questions. Although the technique of interviewing discussed in this chapter deals with listening, the interviewer can obtain considerable supplementary information by visually observing the respondents and their environments.

Data gathered by asking for information is filtered through two humans—the researcher and the informant—and is additionally subjected to the filtering that occurs in the communication process itself. The filtering that results from human perceptions and abstractions is discussed elsewhere (Chapter 1). In this chapter the focus is on the pitfalls that cause inaccurate and incomplete communication between persons and hence distorts the data. Included in this chapter are answers to the following questions:

In what circumstances must researchers depend on data provided by respondents?
What are some techniques for gaining information by asking questions orally?
What are some techniques for gaining information by asking questions in a written form?
What are some advantages and disadvantages of asking questions through an interview, a schedule, and a questionnaire?

Needs for Data Supplied by Informants

The techniques that rely on informants to impart knowledge are important in at least two (or three) different circumstances. These are discussed here under the following headings: data lacking locational precision and the absence of recorded data.

Data Lacking Locational Precision

A frustrating situation commonly faced by geographers is to find data already available but without adequate locational information. The locational facts are not given by the minimum areal unit decided by the researcher but instead by census tracts, counties, or other larger areas. Many characteristics of population (e.g., number of persons per age category, median income, level of education) are obtained by and published in government censuses, but these census statistics reveal only totals and averages for large areal units. A reader is not able to obtain social facts about particular individuals within an enumeration district nor about the exact locations of the individuals. The aggregation of census information by areal subdivisions greatly restricts the use of census data for many geographic problems.

Ms. S., for example, can look in a recent publication of the U.S. Census of Population and find figures tabled for the number of persons living in Choros who are of elementary age. Because the entire city consists of a single census enumeration district, these data have little value for her when she attempts to calculate a point of minimum aggregate travel within the city. She must have a much more precise residential location for each elementary child. Stated otherwise, she cannot differentiate among sites within her study area because the published data treat the census district as a homogeneous "place" with no internal variation.

The frustration that Ms. S. experiences with inadequate census data is a very typical illustration. Geographers and other social scientists who study relationships among human populations are often thwarted by the vast quantities of census data that are available but unusable because of their broad areal grouping. In spite of this limitation, scholars sometimes must accept the census data in their search for relationships, even though they would prefer a different scale for their studies. Such a compromise is unfortunate because of the enormous impact that scale has on spatial associations.[1] The alternative to accepting an inappropriate scale is to collect data that are keyed to locations more precisely than given for census tracts. To collect such data often means asking people to supply the facts.

Absence of Recorded Data

The second situation that requires acquiring field data arises when the researcher needs facts that have not been collected at all. The demand for societal information that goes beyond census data is very common, as attested by the numerous public polls that are taken and reported frequently on a wide range of topics. Even if the scale of study is such that areal units the size of census tracts are appropriate, a researcher may still need original field data not gathered already that can be combined with census facts. For instance, the government censuses contain virtually no information about attitudes, beliefs, or similar personal attributes, which means geographers must often collect at least some of the data needed to solve research problems.

The absence of recorded data might be regarded as encompassing two slightly different situations. One refers to the need for contemporary data; the other concerns acquiring facts about the past. The two situations are similar because they both involve data not available in records, but the one using historical data must depend more on the ability of respondents to recall facts. Techniques that rely on recall may experience difficulties in addition to the usual ones connected with gathering social data because of the filtering of details that occurs between the date of the historical happenings and the time when a respondent is providing information to the researcher. Nevertheless, asking questions of and listening to informants is an important method of acquiring many kinds of historical data.

The methods of obtaining verbalized information vary in several ways such as the form of questioning and the form of answering. Here they are grouped according to the amount of researcher control under the headings of interview, direct schedules, telephone schedules, and questionnaires.

Interviews

An interview is a purposeful conversation conducted in a disciplined way. It is not "just talking to people", even though an interview has all the characteristics of normal, sociable conversation between persons. An interview is more than a two-way conversation because the participants in the dialogue assume the roles of interrogator and informant. Some subsidiary conversation may occur during the interview period and be classed as informal talk between equals, but the core of the verbal exchange is between the one participant as information-seeker and the other participant as respondent.[2]

The term interviewing usually applies to a variety of techniques designed to elicit information from respondents. This variation makes it difficult to specify strategies that apply to all the many forms of questioning and responding. Conversely, because the variations tend to be transitional, suggestions for utilizing a particular subtype may not be restricted to just that one but may be partially applicable to other forms.

Various Types of Interviews

One way of dealing with the large diversity in interviewing is to classify this general technique according to the amount of control exercised by the researcher. In some situations an interviewer may choose to adopt a passive role that resembles that of a participant observer. In others, a researcher may decide on the opposite extreme by exercising tight control in a highly structured interview. Here interviews are classified into four groups: (1) the unstructured interview, (2) the free story, (3) the structured interview, and (4) the structured interview with aids.

Unstructured Interviews

The interview type that most closely approximates a normal converation may be termed an *unstructured interview.* This style is used by researchers who want to maintain a naturalness in the communication process. In many respects the interview may resemble the exchanges that are initiated by a participant observer, except for the fact that the term interview normally refers to a situation where the interviewee is aware of being queried by a person who is asking questions for a particular purpose.

Because the strategy in an unstructured interview is to allow the conversation to develop in the same manner as do most social conversations, the topics may wander away from the principal questions the researcher has in mind. Even though the dialogue may seem unstructured to a person other than the interviewer, in reality the researcher should have a definite set of data goals planned for the interview. If the comments wander too far from the primary goal, the researcher needs to guide the conversation gently back to the desired topic. Inquiries about various subtopics can be woven into the discussion where they seem to fit best. In fact, the skilled interviewer, as exemplified by many TV talk show hosts, can obtain considerable information without the respondent being aware of giving answers to a pre-determined set of questions.

Although the unstructured interview is described here as a technique for gathering data about a respondent's personal views, the same style of questioning can be used for other purposes. One common reason is to seek information that respondents may know about phenomena other than themselves. This goal is especially useful during the time of field reconnaissance, but it may also be an objective when gathering information indirectly (Chapter 8).

Free Stories

A variant form of the non-structured interview is the *free story* approach. This type of interviewing commences with the researcher requesting information on a specified topic, but then allowing the respondent to talk at some length about various aspects of that topic. In this way the interviewer establishes the focus or subject for the conversation, but the respondent has immense freedom in the way the information is expressed.

Free story interviewing refers to the manner of conveying information from the respondent to the researcher—it does not specify the kind of information conveyed. When this type of interviewing is employed to extract a restricted set of facts, it may be known by other names. For instance, when the story deals with the respondent's own past, then it may be called a *life history*. Because normally the data generated from a life history is similar to that obtained by archival research, this technique is explained in Chapter 8.

The free story method is also used in a special way to acquire information about the way respondents perceive their natural environment. Although the free story response is often generated without supplemental aids, sometimes respondents can focus on an issue better if they talk about a tangible object. One way of initiating a story is to present a picture and then ask the respondent to describe what the picture is showing. When this technique is used to learn about the perceptions people have of natural environments and environmental hazards, it is sometimes called *environmental apperception testing*.[3]

Structured Interviews

In most respects the style that best typifies the interviewing technique is the *structured interview*. It is a verbal exchange in which the researcher controls the conversation through an orderly set of specific questions. Outwardly the researcher may not display any more evidence of a specific goal than in the less formal forms of interviewing, but inwardly the interviewer is mindful of a set of questions that will be asked in a prescribed sequence. In order to gain greater objectivity in responses the interviewer may sacrifice some of the contextual naturalness that can be achieved under the unstructured exchanges.

The major theme of this chapter concerns measuring the attributes of human individuals by having respondents express their own attitudes, beliefs, and feelings. Interviewing techniques are also utilized to gather other kinds of data such as group information and group views. When interviewing a group of persons assembled in one place, the interviewer must be able to lead a group discussion in a manner that encourages participation by all persons. Guiding the group responses may not be too difficult when the goal is to gain information about phenomena other than participants' own views. When the objective is to obtain personal feelings, however, structuring the exchanges by group members is more critical. One strategy for acquiring opinions that represent the group as a whole is an interactive procedure often called the *Delphi method*.[4] A primary purpose of this method is to move toward a consensus by having members alternately express their own views and then re-considering them in the light of major themes proposed by the group. This strategy may tend to alter individual viewpoints, but it may also aid some individuals in formulating and defining their personal views.

Structured Interviews with Aids

Even greater control in directing the responses of an interviewee may be achieved in a structured interview with aids. The aids may be maps, pictures, toy models, and similar tangible aids, other than a set of questions printed as a schedule (see below).[5] Aids direct attention to designated topics without requiring the researcher to anticipate all the many variations in possible replies (see "Wording of Response Choices" below). Also, the respondent may feel less inhibited in revealing personal views if such are expressed indirectly through toys and similar objects (see comments on projective techniques below). Sometimes aids can be helpful when a researcher attempts to obtain information from children or persons who are not fluent in the field worker's language.

Having respondents perform a definite task or activity (other than answering a questionnaire) can be done in a variety of ways for many kinds of data. Two techniques typifying the diversity are perceptual mapping and Q-sorting. The former normally commences with little more than a pencil and a sheet of paper on which each interviewee is asked to sketch specified elements in an environment.[6] Preparing and explaining instructions are often difficult because the instructions must structure each respondent's spatial images into a form that can be manifested as a sketch-map.

The second illustration of a structured interview with aids is a specialized technique called Q-sorting. It requires respondents to sort a large number of cards, either written or pictorial ones, into several categories. The purpose is to obtain data about respondents' veiws of classificatory criteria and relationships among the elements represented by the cards. The goal resembles establishing a scaled set of statements by a researcher (Chapter 3) but differs in that the aim is to have respondents reveal their own dimensions for scaling phenomena. In contrast to perceptual mapping, the implementation of this technique depends less on the researcher's verbal guidance than on the preparation of the aids, i.e., the numerous cards to be classified.

With this great variety of interviewing forms and goals, it is difficult to specify procedures that pertain equally to all types. Most of the following suggestions, therefore, apply primarily to the type identified as the structured interview.

Preparing for the Interview

A primary task prior to conducting an interview is to think about potential responses. Just having in mind some questions about the research phenomenon is insufficient preparation for the prospective interviewer. The researcher must also predict and contemplate the kind of replies that may occur so appropriate follow-up questions may bring out more complete information. The necessity for anticipating possible responses to questions is very apparent when one designs a questionnaire (see below). The benefits of preparing for an interviewee's replies is equally important.

Suppose Ms. S. wants to obtain data from residents in Choros about their views on the importance of distance from school. She decides to conduct a structured interview with a sample of residents. Without giving much thought to possible responses, she might choose to ask each interviewee, "Do you think the distance you live from the school building is important?" At first glance this appears to be the question that will give her the information she wants. However, consider some probable replies. A reply of "yes" indicates that the respondent agrees that distance is important, but it provides no measure of the intensity of agreement for varying distances. Maybe she could follow a "yes" response with the question, "How important is it to you?" What kind of reply would this produce? What does a reply of "somewhat important" really mean when translated into a factor affecting a school building location? What does a negative answer to the original question mean? Are there any circumstances when the resident might wish to live closer or farther from the school? Should she be satisfied with a "don't know" reply? If not, what follow-up question could she pose?

Consideration by Ms. S. of these response ambiguities might lead to an alternative strategy. She might try presenting some choices that would provide comparative data and, hence, indicate the degree of importance with which residents regard the issue. Her questioning might proceed something like the following: "Does your child walk to school?" "If so, how far does he/she walk?" "If you moved to another house, how far would you be willing to have your child walk to school?" "If you take your child in a car, how far do you drive to school?" "Would your child walk if you lived closer to school?" "If you lived farther from school, would you prefer having your child ride a school bus?" "How far would you be willing to drive your child to school instead of sending him/her by school bus?" These choices demonstrate the use of questions that will measure responses by forcing comparisons and indicating priorities. Concern here, however, is less with the particular type of interview questions than with the fact that Ms. S. must try to anticipate interviewee responses. If she ponders on the probable replies to her impulsive set of questions, she would undoubtedly realize the need to revise her line of questioning.

A researcher should also give prior attention to the subsidiary information available in an interview situation. Although an interview consists primarily of oral communication, its setting provides the potential for obtaining relevant supplementary data. A skilled researcher can gain much insight from the respondent's non-verbal communication, such as (1) the use of gestures of hands, facial expressions, and body positions (kinesics); (2) the use of volume, pitch, and voice quality (paralinguistics); (3) the rate of speaking (chronemics); and (4) the use of interpersonal space (proxemics). These auxiliary communicators will not necessarily answer specific questions posed by the researcher, but they often reveal clues about the authenticity and reliability of the informant's replies. In fact, meaningful data are highly dependent on the researcher's assessment of the informant as a reliable source of data. This means that an experienced interviewer should be able to evaluate the various factors that inhibit an informant from communicating accurate data. In many respects, this ability to assess the respondent's answers is essentially the art of listening perceptively. "Often the neophyte interviewer is deceived into thinking that the interview has been productive because the respondent has 'talked on and on.' He may accept this as evidence that no inhibitors were at work in the situation. Unfortunately, this is not necessarily true."[7]

Another preparatory consideration is the timing of interviews. Without adequate thought given to the time at which interviews will be conducted, bias or distortions may result in at least two ways. One way concerns the sampling frame. Even if the locations of the residences where interviewing is to be conducted have been selected randomly, a rather blatant bias would undoubtedly occur if interviews were conducted with only "the first adult who answered the door" when contacts were made during the midmorning on week days. At least those persons might represent a different population than those who answered the door on Sunday afternoons. This aspect of timing will not be disadvantageous if the sampling frame is a list of employees, members of a social organization, or another frame not related to which persons are in particular places at specific times.

A second and more general form of distortion may result from timing of the interview because the hour, day, or season may affect the receptiveness of respondents and, consequently, their answers. Attitudinal data collected from relaxed informants, for instance, may differ from that given by busy persons who are anxious about the time consumed by the interview. Therefore, some preparatory thoughts should be devoted to choosing appropriate times for interviewing a particular target population.

Also, prior to any field interviewing, researchers should prepare to satisfy any challenge to the legitimacy of their investigations. In other words, researchers must clarify in their own minds the reasons for interrupting the lives of others and for requesting the time and cooperation of potential respondents. The merits of gaining knowledge that will benefit society must be weighed against the interruptions to individuals' private lives (see "Ethical Issue of Collecting Data," Table 5.2). Furthermore, each field worker should carry a letter of introduction that verifies the credentials of the interviewer and states the purpose of the project. This letter should also give assurance that the rights and welfare of the respondents are adequately protected with respect to security, privacy, confidentiality, embarrassment, discomfort, and harassment. In essence, it declares that the risks to the informants are outweighed by the expected benefits to society.

Approaching the Interview

No high-pressured gimmicks will guarantee entrance to a stranger's home nor assure participation in an interview. An amateur interviewer can increase the rate of cooperation, however, by reducing potential confusion in the minds of prospective interviewees. The immediate objective of the interviewer is to separate the beneficial purpose of the visit from the sometimes more annoying or threatening behavior of investigators, salespersons, and molesters. The art of communicating this distinction depends on a multitude of variable conditions, so it is difficult to prescribe specific dress, wording, attitudes, gestures, or other behavioral manners. In general, the field worker should refrain from moving in an aggressive manner, should avoid artificiality in speech, and should shun the word "investigation". Giving one's name to the respondent in the introduction may be an aid in reducing suspicion, as well as being a common courtesy.

The contents of a successful introduction usually include: (1) a concise statement about the purpose of the study, (2) the identity of the sponsor or agency conducting the research, (3) a verbal affirmation about the confidentiality of any volunteered information, and (4) a brief description of how the respondent was selected. If, after the first introduction, the potential interviewee does not reply affirmatively to a request for cooperation, the researcher must be ready to provide more details. To questions about legitimacy, the researcher should display a letter of introduction and expand on its contents. To comments about being rather busy, the interviewer needs to provide a prompt statement about the expected length of time for the interview. Also, the researcher should be prepared to schedule a future time for interviewing if the respondent is, indeed, too busy. In general, the interviewer should manifest confidence in the value of the project without exerting obnoxious zeal. This means the field worker should not suggest something that might give prospective interviewees a reason for non-cooperation (e.g., "Are you too busy?").

The procedures described in the preceding paragraph are usually applicable for approaching the so-called middle class population of American society when the researcher wants to gain information through an interview, schedule, or questionnaire. The approach may need to be modified if the researcher is requesting cooperation from persons in different circumstances. Special strategies may be appropriate when prospective respondents shun publicity (e.g., occupants in an illegal gambling hall), have distinctive characteristics such as being famous, or are from a different culture. In these cases the approach may have to be made through the intervention of influential members of the population to which the prospective interviewees belong.

Questioning

Obtaining valid information from respondents depends on the researcher's experience, intuition or common sense, and understanding of respondents' perspectives.[8] This understanding is essentially the realization that certain conditions may cause interviewees to be unable or unwilling to give accurate information. Prospective informants may be unable to provide certain facts because they do not know, have forgotten, or are confused about them. They may be unwilling to give complete or accurate information because they feel threatened, embarrassed, inhibited, or uninterested. Through experience and common sense an interviewer can detect some of these inhibitors and can diminish their detrimental effects. The interviewer can also utilize various rewards to counteract the inhibiting factors. Rewards may be in the form of contributing to a new experience, offering companionship, providing an outlet for expressing personal beliefs and feelings, giving recognition, and supplying sympathy.

The atmosphere the interviewer should promote is one of appreciation for the interviewee's contributions. Appreciation can be expressed through attentiveness and encouraging words and gestures. Because the purpose is to obtain the views of the respondent, the researcher should refrain from displaying shock at or approval/disapproval of particular responses. Similarly, the objective stance of the interviewer should prohibit trying to prove inconsistencies or demonstrate weaknesses in the respondent's logic.

Acceptance of the interviewee's honest opinions and viewpoints is not equivalent to accepting all statements uncritically. If the interviewer suspects that the respondent is giving deceptive or invalid replies, the interviewer should consider alternate ways of asking for information in order to gain the confidence of the interviewee.[9] Awareness of nonverbal clues should aid the researcher in assessing the degree to which cooperation exists. If the responses seem to be honest ones but contradictory to some degree, then the researcher should attempt to clarify inconsistencies in an inquisitive manner that does not accuse the informant of deliberate misinformation but rather guides the respondent toward revised answers. Often a summarizing review of major points at the end of the interview is helpful in clarifying answers.

Closing the Interview

Although more preparatory attention is given to commencing an interview than closing it, the latter should not be neglected. Planning should take into consideration the concluding atmosphere and ways of terminating the interviewing session. The personal relations between the researcher and respondent at the end of an interview should be as cordial as at the beginning. Ideally, every interview ends with the respondent feeling good about the exchange and about having contributed to the study.

Occasionally interviewees experience so much satisfaction in the exchange that they are reluctant to end conversations. This may be more prevalent among elderly, single persons, but it certainly is not restricted to that population. The interviewer should give prior thought to ways of terminating a prolonged interview (which may consume valuable research time) without damaging the cordiality of the conclusion.

Recording Information

The communication process involves not only the questioning and answering, which constitute the core of the interview itself, but also listening. It is what the interviewer hears and records that becomes the data used in subsequent analyses. Therefore, the researcher must utilize procedures that will insure fidelity in listening and recording. Although there are no standard procedures for listening to interviews because of the many variable conditions, the researcher should not ignore the importance of this phase of the interview. "The art of 'scientific' listening is. . .the most neglected technique in the researcher's field kit."[10]

That the researcher truly listens is of foremost importance. Scientific listening differs from social conversations where persons often think more about what they themselves are saying or planning to say than on the contents of what other speakers are saying. In some interviews this listening component may require some special background reading and/or experience from a pilot project to prepare the interviewer for fully understanding what is being said by the informant. Of course the researcher should be informed about the research topic in general, but preparing to listen for and to understand particular expressions of respondents may require additional forethought. Also, to be a good listener frequently means having patience—patience to hear rambling responses and imprecise discourses without outwardly displaying impatience.

Methods for recording what is heard vary according to the formality of the interview, the environment in which it occurs, the kind of data the researcher is seeking, and the wishes of the respondent. Data may be memorized and written later, reported by notes during the interview, or recorded mechanically by a tape recorder.

Researchers who choose to memorize information long enough to write the results after the interview often do so because they believe the act of recording will interfere with the rapport or is inappropriate. Memorization is common when the form of the interview is unstructured and the precise wording of responses is not required. By concentrating on key words, the researcher is able to remember the contents of the interview until after it is concluded. As soon after the interview as possible, the researcher writes notes on what was said, and these notes become the data for subsequent analysis.

Note-taking is a very common method for recording remarks made during an interview. The respondent/informant knows the interviewer is seeking information and expects such facts to be written. Obviously the note-taking should not become so tedious that it interferes with a smooth flow of communication or greatly prolongs the interview time. Using abbreviations and coding symbols in the same manner as codes are used for visual classifications is helpful in shortening the interview time and organizing the response data. Sometimes repeating some of the respondent's phrases orally gives the researcher a little more time for writing without causing silent pauses. This also may reassure the interviewee that his/her views are being recorded correctly. Whether or not these notes are rewritten after the interview into a more legible or complete form depends on the original recording format and the kind of data being collected. If the researcher is seeking responses that were roughly categorized prior to the interview, coded notes may be adequate without additional transcribing.

In other circumstances the most advantageous method of reporting an interview may be by using a tape recorder. Some kinds of data are complex enough that analysis at the time of the interview is almost impossible. Linguistic relationships among verbalized ideas (e.g., content analysis; Chapter 8) illustrates data that are too difficult to collect without having them recorded for later replay. Having records of interviews obtained for less complex analyses may be useful too, partly as a check on the objectivity of extracting data from the interview.

At the beginning of an interview the researcher ought to explain reasons and secure permission for using a tape recorder. The equipment, which should have been previously prepared for operation with only the push of a button, can then be left in an inconspicuous place and generally ignored by the interviewer.

The advantages of a complete record of an interview are not achieved without cost. One cost is the monetary expense of the equipment. Another is the tremendous amount of time it takes to transcribe the recorded materials. A third cost is the potential reluctance of respondents to express personal feelings and opinions for what they perceive could be a permanent record. Such reticence may vary from an outright refusal to cooperate if a tape recorder is used to a more subtle restraint in sharing personal information. The researcher must certainly be sensitive to these subtler forms of data modification when assessing the merits of using a mechanical recorder in an interview.

Schedules

A schedule which is one kind of instrument designed to gather information from respondents, consists of a written or printed list of questions that are to be asked by the researcher and a set of accompanying choices of responses to those questions. A direct contact schedule is very similar to a structured interview using aids (see above), but in this case the supplementary "aid" is a sheet or booklet of questions. It is also similar to a generally structured interview except the researcher records responses on a previously prepared code sheet. Furthermore, in some cases there may not be a clear cut division between a schedule and a questionnaire (in fact, some researchers use the terms interchangeably) because both require the informant to respond to a written set of questions. In this text, however, the term questionnaire usually pertains to a set of questions that is self-administered. Many of the following comments about the wording of questions are appropriate for both schedules and questionnaires because the forms of questions and the choices of responses are similar.[11]

Although the overall objective for utilizing a schedule is similar to the goal of interviewing— the collection of data by asking persons to volunteer information—the procedures differ in the form of responses. Whereas the interview is designed to obtain information in the terminology and expressions the respondent chooses, the structure of the schedule usually forces respondents to accept the wording of the researcher. As a result, the nature of the data tends to differ also. Interview responses provide a wide variety of data, ranging from the content of answers to the manner in which the answers are communicated. Schedules frequently produce more quantifiable data. Moreover, a schedule's greater dependence on the wording and choices of responses places more emphasis on preparing meaningful questions and valid response options. In an interview the questioning and answering can continue until the researcher is satisfied that a valid interpretation of the respondent's intended reply has been conveyed. In a schedule, the wording of the questions and the potential responses must transmit most of the message without further assistance.

Even if researchers expect to use a schedule prepared by someone else, it is important that they understand fully the implications of the wording of the schedule. More likely than not, no existing schedule will provide satisfactorily the particular data needed by an individual researcher. In these circumstances an original schedule will have to be created, and the researcher is obligated to construct an instrument for meaningful communication. Many people naively believe that a valid questionnaire or schedule can be assembled by writing down a few questions. Unfortunately, composing a set of questions and response choices that will accurately communicate valid data is a complex task. Writing a useful schedule or questionnaire requires careful attention to many details.[12]

Background Data

The easiest portion of a schedule to design is normally the introductory part which provides information about the research project and requests background data. It commonly begins with a title, the name of the sponsoring agency, and a statement of confidentiality. The latter might resemble the following:

The information obtained will be used for statistical purposes only. The confidentiality of the individual informant's identity and answers will be respected.

Near the beginning there should be blank lines where the researcher can write the following facts: (1) an I.D. number, (2) the location of the interviewee, (3) the date and time of the interview/schedule, and (4) the field worker's name.

Census-type information about the respondent is normally positioned near the beginning of most schedules, although sometimes it is at the end. Any social variable that is hypothesized to be related to the problem phenomenon needs to be recorded, such as age, sex, marital status, and level of schooling. Conversely, there is no reason to include factors that are not expected to be a part of the research. Every additional question lengthens the schedule and the time necessary for completing it. Some facts may be acquired through observations and, hence, can bypass the questioning procedure. For instance, sex, approximate age group, and environmental conditions typify questions that can be observed visually by the researcher without consuming interview time.

Wording of Questions and Responses

The writer of a schedule must be concerned about both the wording of the questions and the sets of responses. The questions and response choices are not unrelated, but they do possess somewhat different communication tasks for the writer of a schedule. Each question should be worded to reflect some aspect of the research goal. The response choices, except for open-ended questions, represent the researcher's prediction of the respondents' replies. The wording of these choices should provide the informants the opportunity to provide accurate information by choosing from among a set of unambiguous options.

Table 6.1   Specificity of Questions for a Schedule or Questionnaire

a. Classes

Level of Specificity	Core Terms				
	(AR) Actor	(AN) Action	(RE) Relationship	(WR) Where	(WN) When
I Most Specific	the youngest child of elementary age in this family (the youngest)	walk	with other children	from home to school	last Monday morning
II	any children in this family	travel	with anyone	in Choros	usually
III Least Specific	anyone in this family	–	–	–	–

b. Some typical questions classified according to their combinations of the core terms.

　　　AR-I*　　　　AN-I　　　　RE-I　　　　　　WR-I　　　　　　WN-I
1. Did the youngest** walk with other children from home to school last Monday morning?

　　　AR-I　　　WN-II　AN-I　　　　RE-I　　　WR-I
2. Does the youngest usually walk with other children from home to school?

　　　AR-I　　　WN-II　AN-I　　　　RE-I　　　WR-II
3. Does the youngest usually walk with other children in Choros?

　　　AR-I　　　WN-II　AN-I　　　RE-II　　　WR-II
4. Does the youngest usually walk with anyone else in Choros?

　　　AR-I　　　WN-II　AN-II　　　RE-II　　　WR-II
5. Does the youngest usually travel with anyone else in Choros?

　　　AR-I　　　　AN-I　　　　RE-I　　　WR-II　　　　　　WN-III
6. Does the youngest walk with other children in Choros?　　　　　　　——

　　　AR-I　　　　AN-I　　　　RE-I　　　　　　　WR-III　WN-III
7. Does the youngest walk with other children?　　　　　　　　　——　　——

　　　AR-I　　　　AN-I　　　WR-I　　　　RE-III　WN-III
8. Does the youngest walk from home to school?　　　　——　　——

　　　AR-I　　　　AN-I　　　　　　　　RE-III　WR-III　WN-III
9. Does the youngest walk?　　　　　　　　　　　——　　——　　——

　　　AR-II　　　　　　　AN-I　　　　RE-III　WR-III　WN-III
10. Do any children in this family walk?　　　——　　——　　——

　　　AR-III　　　　　　AN-II　　　　RE-III　WR-III　WN-III
11. Does anyone in this family travel?　　　　——　　——　　——

* Class indicated in Part a

** Abbreviated here for "the youngest child of elementary age in this family"

Table 6.2 Two Surveys about One Issue with Contrasting Wording (By the Selection Research Inc. of Lincoln for the *Lincoln Sunday Journal and Star,* published December 19, 1971)

a. Survey of November 1971:

Question	Responses (Percentages)
Which proposal do you favor?	
Drop the present sales tax on food with the $7 refund	57
Retain the present sales tax and refund	39
No opinion	4

b. Survey of December 1971:

Question	Responses (Percentages)
The proposed exemption of food from the sales tax would deprive Nebraska state government of an estimated five million dollars a year in revenue. In view of that prospect, which action would you prefer to see taken?	
Make up the lost revenue through higher tax rates	16
Not make up the loss but reduce state services	23
Not exempt food from the sales tax	59
No opinion	2

Wording of Questions

The wording of questions may vary according to their specificity of actor, action, relationships, location, and timing. Three levels of specificity with several combinations of these levels can illustrate the framework of various questions that might deal with a single topic (Table 6.1). These variations in what is essentially the same question may occur because researchers differ in the extent of control they want to impose on the responses. In general, the less specific questions (Level III in Table 6.1.a) are best related to response forms that are open-ended. Otherwise, the lack of specificity combined with rigidity defined response choices will cause ambiguity, thereby reducing the reliability of replies. Using Table 6.1.b as an example, if respondents were restricted to only a "yes" or "no" reply for Question #9, the researcher might wonder if everyone interpreted the question in its general and literal meaning or whether it was interpreted to mean the same as Question #8.

In contrast to the fault of wording questions too vaguely, other authors may be inclined to present too much and/or a one-sided explanatory background when writing questions. It is usually necessary to provide some preliminary explanation about the kind of information being requested, but the writer must be extremely careful that the background statements do not influence respondents to change their answers. The influence of an introductory statement is illustrated by two surveys taken within a month of each other in Nebraska (Table 6.2). In November only 39% of the sampled respondents favored retaining the sales tax on food, but in December 59% of the population preferred the retention of the tax on food. Although it is theoretically possible that the sampled population did make a dramatic change in one month and/or that the samples are extreme aberrations, most analysts would conclude that the different results reflect the wording of the questions and the specific set of response choices.

Praseology that suggests a particular response must be avoided when the goal is to determine the respondents' true attitudes, beliefs, and feelings. This is not to deny that some research goals may require the kind of data that are produced by emotional or suggestive questions, but under normal circumstances writers of scholarly schedules should avoid language that suggests a particular response. Unfortunately, schedules are sometimes misused to create the impression of popular support for a particular position rather than to ascertain objectively a population's opinions. A common tactic for obtaining misleading results is to word questions so they suggest only one "sensible" answer (Table 6.3).

The inclusion of influential materials may assist in producing meaningful data in special circumstances. The interactive strategies of the Delphi method is a special case that depends on respondents knowing the first-round replies of each other. The strategy is to provide respondents intentionally with this information for the purpose of letting them modify their responses on subsequent rounds. This specialized technique, however, is not used by most researchers. In general, the admonition about avoiding words and phrases that may cause respondents to alter their replies to schedules and questionnaires still applies.

The linguistic style of individual questions is critical for effective communication. In general, vocabulary should be simple and not laced with esoteric jargon, double negatives, and ambiguous terms. Sentences must be short and precise. At times, misleading data result from schedules containing multiple sentences or questions with multiple interpretations. For example, consider the question: "Should Choros build one new elementary school building and locate it at the same site as the present building?" A citizen might give a negative reply because (1) the respondent prefers constructing two buildings, or even no building, rather than one, or (2) the respondent agrees with having one new facility but does not want it re-built at the present school site. This kind of error is committed very frequently by inexperienced questioners who inadvertently combine two questions in a single sentence.

One strategy for encouraging respondents to give freer and less inhibited responses to very personal inquiries is to word questions in the third person. This enables informants to "project" their own opinions onto others rather than having to reveal their thoughts directly. Respondents can be directed to provide their answers in a variety of forms, so the set of different ways are known collectively as *projective techniques* (with the environmental apperception testing through interviewing being one version). A typical question is expressed as follows: "In your opinion, why do some people in Choros prefer renovating the elementary school building rather than constructing

Table 6.3   Wording that Suggests a Particular Response (By Congressman X, January 1968)

Question	Response Choices
Do you favor expanding trade with Russia and other communist countries, even though they continue to supply our enemies in Vietnam?	Yes No Undecided

new facilities?" Questions may be matched with a variety of different response types, but the commonality in projective techniques is the indirect manner in which respondents are asked to provide information about themselves.

Wording of Response Choices

In many respects, the writing of response choices contains even greater pitfalls for the amateur researcher than the wording of questions. Except for the response form of open-ended questions, the informant is given a choice of responses which are ones the researcher has previously considered and written. This task requires the author of a schedule to be familiar enough with the phenomenon and the attributes of the respondents to be able to predict all their potential replies. One of the major skills in writing response choices, therefore, is to be able to determine whether each set of stated options is completely inclusive.

Another major consideration in preparing sets of responses is the exclusiveness of choices. In essence, the informants are asked to measure some phenomenon (such as their attitude toward walking or their spatial behavior last week) by classifying it. Consequently, the set of classes must not be overlapping and must not permit a single individual to belong to more than one class.

Ways of avoiding these two potential difficulties, namely, not being inclusive and not having mutually exclusive classes, vary somewhat with the different forms of responses. Responses that present the informant with a dichotomous choice are useful when the requested information can be easily polarized (e.g., the respondent is or is not currently a resident of Choros; Table 3.2, #1). Respondent replies are uncomplicated and thus simple to record on the schedule form. Difficulties may occur when the respondent's perspective is not easily polarized. In such cases the either-or type of choice forces a misleading reply. For example, a question that inquires about the suitability of the present location of the Choros elementary school may actually tap affirmative opinions ranging from "It's the *only* logical place" to "Oh, it's okay". All of these replies would be lumped together as "Yes" on a dichotomous form. In other circumstances it may be appropriate to force a dichotomous choice, for instance, to predict the outcome of an election. It is the researcher's responsibility to decide which of these conditions applies to a particular research goal.

A companion dilemma facing the researcher who writes a dichotomous choice is whether or not to provide for a reply of "Don't know" or "No opinion". Normally it is better to include such a response choice because it allows informants to indicate their lack of preference. One reason this is important is because there is a general tendency for many people to give a reply, even an arbitrary one, to questions rather than admit ignorance. The absence of a "No opinion" choice only accentuates this tendency. Another reason is that a large percentage of "Don't know/No opinion" responses to a specific question indicates to the researcher that the question is ineffective. Without including such a choice, the researcher may not know that respondents, in fact, were confused about the question but marked an arbitrary answer anyway.

The multiple choice form of response that presents the informant with a set of response categories overcomes at least one of the limitations of the dichotomous choice. Multiple choices do not force what is in reality a wide variation of opinions among the sampled population into only two measurement classes. Writing multiple choice responses is, however, difficult for many aspiring questioners because they may fail to make the choices both inclusive and mutually exclusive. Illustrations of this kind of failure are numerous, but two questions that appeared in a newspaper survey demonstrate the point (Table 6.4). It would appear from the results that in

Table 6.4   Illustrations of Nonmutually Exclusive Response Choices (By the *Lincoln Sunday Journal and Star,* September 5, 1971)

Questions and Choices	Responses (Percentages)
1. Today's children would be better behaved if:	
a) Home discipline were stronger	69
b) Home discipline were more lenient	10
c) Fathers took more responsibility for discipline	18
d) Mothers stayed home more frequently	3
2. Sexual relations are acceptable:	
a) For married women	25
b) For unmarried women	21
c) Never outside of marriage	54

Lincoln, Nebraska, three-fourths or more of the respondents (1) do not believe children would behave better if mothers stayed home more and fathers took more responsibility for discipline, and (2) do adhere to a belief that married women should practice complete sexual abstinence. Most interpreters doubt the validity of these measurements, however, and would insist that the overlapping choices (classes) caused the surprising results.

When choices are numerous or lengthy, it is wise for the researcher to write each choice on a separate card. At the time the schedule is administered, the researcher can hand the set of cards (choices) for a particular question to the informant for reading, comparing, and making a selection. Having the choices on cards has the added advantage that, if they are shuffled after each informant's reply, any bias that might occur because of order of choices is diminished. This strategy, of course, depends on the respondent being able to read, but this literary factor is not a handicap for many target populations. Also, cards with pictures can be used to depict choices among certain phenomena.

Placing each choice on a separate card or sheet is especially helpful when obtaining data through structured scenarios and card sorting. *Scenarios* are descriptions of a few alternative outcomes predicted for a chain of events, described in the core of the question (Table 6.5).[13] Although the researcher can read each scenario choice to the respondent, sometimes the descriptions of the possible outcomes (scenarios) are complex enough that many respondents prefer reading and re-reading the various alternatives. Having each scenario on a separate sheet facilitates this preference.

Card sorting requires respondents to arrange numerous terms or statements into rankings or other prescribed classes.[14] Without using cards, the researcher must recite through the entire list of terms or statements so the respondent can conceptualize their range and variance and hence an appropriate set of classes. Each term or statement must then be repeated while the respondent indicates a class answer. The technique is much easier if each term or statement is written on a separate card because this permits the respondent the freedom to rearrange choices while pondering on final answers. After each respondent completes the task the researcher records the results on the schedule in the same manner as if they had been verbalized replies.

Table 6.5   A Set of Scenario Alternatives

Decisions about the elementary school building that the citizens of Choros make now will affect the city for many years. It is unlikely that the people in Choros school district will decide to change the elementary school building(s) greatly in the near future. Therefore, the decision made now must be based on both conditions which now exist and those expected in the near future. The facts about the current needs are well known. Facts about the future are unknown, but citizens must make decisions now based on their estimates about the future needs for elementary school facilities.

Below are different conditions that might exist in Choros in fifteen years. Choose the statement that best describes what *you* think Choros will be like.

A) The population in Choros will grow rapidly and the city will expand into the surrounding land. The financial support for the schools will grow in accordance with the population growth. In fifteen years the city will need at least three elementary schools, which will need to be widely dispersed to serve all parts of the city.

B) The population in Choros will grow at a moderate rate but the land area will change by only modest amounts. The financial support for the schools will grow in accordance with population growth. In fifteen years the city will need two schools, which should be located in two diverse areas of the current city limits.

C) The elementary school needs in Choros will be very similar in fifteen years to those of the present. Neither the population nor areal extent of the city will change much, and the financial support will remain about the same as now. Whatever school building(s) is/are constructed now will be adequate then.

D) The population in Choros will not change much in fifteen years, but the financial support for the schools will decline relative to the current economic backing. All school facilities, which will then include only one elementary school building will need to be used at maximum capacity and efficiency. Any unused buildings and/or property will be sold.

E) Choros will decline. In fifteen years the population will be smaller and the financial support will decrease in accordance with this decline. One school building will be adequate for all elementary students in the city.

Another form of response that has some of the characteristics of both dichotomous choices and multiple choices is a check list. It may appear similar to the multiple-choice form, but the informant is instructed to choose as many responses as apply. Not being restricted to one response item may be an advantage when dealing with complex ideas but it may also be a disadvantage when the researcher wants to force a fine differentiation between choices. This is because the respondent is not forced to choose among a set of classes. In essence, the informant is internally asking the question "Does this apply or not?" for each item in the list.

The respondent normally expects the list to be a complete set of options for the stated question, so inclusiveness should be a goal. A frequent fault in preparing a check list is the omission of one or more from all possible options (which means the respondent necessarily provides incomplete information), unless a category called "Other" is included. Using an "Other" class makes the list inclusive, but this catch-all category should not be used as a lazy way of making the list complete. If too many respondents give this reply, the researcher usually receives very little information. If the class is titled "Other; list/name _____" and numerous respondents enunciate such replies, the researcher obtains data mostly from what is effectively an open-ended question. This is not necessarily a disadvantage, but it does change the form of the resulting data.

The forms that rely on Likert-type statements and semantic differentials are usually not troublesome for the researcher who is writing response options. In the case of Likert statements the choices are fairly standardized (see Table 3.2, #11–14). Likewise, when the form requests the respondent to react to Likert statements by marking on a line (Table 3.2, #61), the poles of the line are normally indicated by standardized terms. Semantic differentials are also composed of a pair of polar terms, but selection of semantic opposites requires the author to consider carefully the meanings which respondents may impute to the matched terms.

Pre-Testing

One of the most beneficial activities related to constructing a schedule or questionnaire is the time spent on pre-testing. As stated above, writing a schedule that effectively obtains useful data involves more than jotting down a few questions. Even experienced pollsters find it necessary to make trial runs of various preliminary editions of a schedule to determine which is most valid as an instrument for measuring social attributes. Certainly an amateur researcher should plan to devote many hours to the task of writing and testing a schedule before using it in the field to collect research data.

A typical plan for developing a valid schedule or questionnaire might commence with the writer assembling a first draft, and then submitting it to friends for suggestions. After the researcher is reasonably satisfied with a schedule, it might be duplicated for pre-testing. The pre-test could be administered, possibly during a pilot project, to a group of persons selected to represent as much variation in the target population as is convenient. Although this group should be representative of the total population, they need not be randomly selected. Members of pre-test groups, must of course, be excluded from subsequent populations of respondents.

The researcher can use pre-testing in several ways to construct a better schedule. One direct way is by asking the pre-test respondents about their reactions to questions and about their suggestions for improving the schedule. Another way, which supplements the first one, is to analyze responses in the manner explained previously in Chapter 3. Questions that detect little variation or produce erratic results should be modified or omitted. Another way is to try several different wordings, orders, and formats in the pre-test as a basis for studying the results carefully to discover any difficulties that may be related to particular forms. For example, if numerous respondents answer "I don't know/No opinion" to a particular question, then the researcher should doubt the effectiveness of its present wording or approach. Likewise, the researcher may want to explore the effect of varying the order of choices. During the pre-test the order of the choices should be changed on different editions of the schedule to see if informants tend to choose, for example, the response listed first.

Another variation that can be explored during a pre-test is the use of a *sieve question*. Such a question is one that separates respondents into groups to which certain subsequent questions may or may not apply. For example, suppose Ms. S. wanted to learn about the conditions that exist when parents, whose children normally walk to school, occasionally take their offspring in a car. If she includes a sieve question that inquires about how children normally get to school, she can skip part of the schedule when questioning families that usually take their children in a car (Table 6.6). For families that respond to the sieve question in a way that tells Ms. S. that the children do frequently walk, the sequence of questions will proceed with the inquiries about

---

Table 6.6   An Illustration of a Sieve Question

. . .

#8 Does the youngest child of elementary age in this family usually walk to school?

     Yes   _____ (Continue to Question #9)

     No   _____ (Skip to Question #13)

   #9 On those occasions when the child does not walk to school, does someone in the family normally take him/her to school in a car?

        Yes   _____

        No   _____

. . .

#13 Approximately how long does it usually take for the child to travel to school?

       Less than 5 minutes   _____

       5–20 minutes   _____

       More than 20 minutes   _____

       Don't know   _____

. . .

---

conditions that exist when children occasionally do not walk to school. By trying various sieve questions and branching strategies during the pre-test, a researcher may develop a form that is complete but has the capability of being shortened for respondents with certain characteristics.

The order in which questions appear in a schedule or questionnaire should be logical, easy to follow, and contribute to gaining and maintaining rapport. Usually the easiest questions come first to put the respondent at ease and gain confidence. Embarrassing or emotional questions can be placed in the middle or near the end (but not at the very end) of the schedule so rapport with the respondent is not threatened and so the interaction ends on a positive note. Placing general questions prior to specific ones provides a logical grouping because it funnels responses given to least restrictive questions toward subsequent narrower ones. In some cases, however, the researcher may deliberately divide questions that logically belong together and scatter them throughout a schedule for the purpose of checking on reliability and consistency in responses.

The format should be arranged in a manner that eliminates confusion and errors by the administrator of the schedule and by the researcher who reads and tabulates the results. Branching directions must be very clear so the field worker can skip sections quickly but accurately. Because open-ended questions need plenty of space for answers, the pre-test is the proper time to verify the adequacy of room for writing replies. Every question should have enough space for supplementary notations that the researcher may need to make.

In addition to helping the researcher write a better schedule, pre-testing can also assist in developing skills for administering the schedule. The researcher may discover that because the format of the schedule is confusing in some places, a field worker might easily make mistakes. A pre-test may also tell which of two forms, providing essentially the same information, is the one that respondents prefer answering. In addition, the researcher can obtain an estimate of the time it normally takes to complete a schedule. This is a handy fact to have when initiating requests for cooperation from potential respondents.

Telephone Schedules

A telephone interview, which is a schedule administered over the telephone, resembles regular schedules in that the researcher reads the questions and records replies given by the informant. It differs, however, because the interrogator and respondent are not together in the same setting. Their only contact is through a mechanical instrument and the communication is only auditory. This difference and the general popularity of telephone surveys are reasons for considering telephone schedules as a separate topic.

Advantages

The primary advantages of a telephone schedule are its accessibility and speed combined with low cost. When contacted by telephone potential respondents are usually more accessible in terms of physical proximity and convenience than by other modes of communication. In effect, the respondents are as close to the researcher as they are to a telephone. Those who reside in parts of the country distant from the caller, in isolated regions with poor transportation, and in areas with hostile or restricted environments are easier, at times, to contact than persons in an adjoining room. Furthermore, because telephones occupy such a dominant position in American social contacts, most potential respondents will answer a ringing telephone before opening their mail or continuing a conversation with a stranger at the door.

The advantage of speed is closely associated with that of accessibility. A large number of potential respondents can be called faster by telephone than the same researcher can contact them personally or through an exchange of mail. This means that a researcher who has a problem which requires very current data (e.g., the reactions of voters to an event that occurs shortly before election day) is almost forced to use a telephone schedule. Likewise, information based on respondents' recall often depends on prompt questioning. For example, if Ms. S. wanted to ask mothers about the way their children traveled to school on a selected day, she could expect reasonably accurate recall by the mothers if they were called by telephone the same day or the next day. In contrast, if Ms. S. interviewed the mothers in person and it took three weeks to contact all the mothers in the sample, those parents near the end of the interviewing period would probably have forgotten the facts which they regard as fairly routine and trivial.

The costs of telephoning compare favorably with those of traveling and interviewing. During recent times in the United States the cost of using a telephone has been much less than the cost associated with a person traveling the same distance. Furthermore, the cost of the researcher's own time, and possibly the salaries of field assistants, is much less when using telephone schedules. When comparing the costs of using a telephone with those for mailing questionnaires, the advantage is much less and depends on several variables such as the number of long-distance calls and the number of duplicate mailings (see below).

In addition to these primary advantages of a telephone, there are other conditions that may make this technique for asking questions attractive to a researcher. One is the ease of using a telephone directory (or the combination of a directory with randomization of the last four digits) for a sampling frame.[15] A second potential advantage is that some prospective respondents are more likely to agree to answer questions. This is because the remoteness of the interviewer makes them feel less threatened, and they know they can terminate the conversation at any moment. A third advantage, when compared to mail questionnaires, is that the rate of response is invariably much higher for telephone schedules.

Disadvantages

Even though the telephone schedule possesses several advantages, it also has some severe handicaps, among which is the lack of direct visual contact. Researchers cannot utilize all their sensory faculties to observe the source of data and make visual assessments. Although it is true that this chapter emphasizes techniques which rely more on listening than on making visual observations, the lack of supplementary data acquired visually by the researcher limits the effectiveness of purely auditory techniques.

One sacrifice made by using a telephone is the assurance that the responding voice belongs to the intended informant. The speaker may be another person, such as jokester or one who has recorded a message for a mechanical answering device. Likewise, there is no guarantee that the researcher will learn the correct location of the respondent. Although telephone directories usually publish addresses for listed residential homes inside cities, these may be inadequate (e.g., if randomized dialing reaches an unlisted number), locationally imprecise (especially in rural areas), or inaccurate (for instance, outdated numbers). For some research problems the locational precision may not be critical and the zones indicated by the three prefixed digits of telephone numbers may give adequate areal precision. For large-scale studies when geographic data for every informant is necessary, the lack of certainty about locations injects possible errors.

Without concommitant visual observations the researcher is hampered in acquiring supplemental information about characteristics of the respondent and about the environment in which the interview occurs. The interviewer who contacts the respondent directly knows whether the person is participating in a wild party, consoling a grieving friend, frantically preparing to leave, or relaxing alone. The telephone caller can seldom know about any environmental settings which may affect the informant's responses. Also, much non-verbal information is unavailable to the telephone interviewer. For example, the caller may be less able to assess the degree to which the respondent is irritated by the intrusion than when observing the respondent directly. While pauses, inflections, and similar paralinguistic clues may reveal respondent reactions, other reactive indicators are unknown to the researcher. The lack of these supplementary observations makes it difficult to judge the reliability of many telephone replies.

The lack of visual clues can hamper the researcher also because the prospective respondent is unable to make visual judgments about the characteristics and motivations of the interviewer. Without these visual aids the respondent has to judge the legitimacy of the caller's objectives on the basis of the first few sentences. Because crank calls and a tremendous amount of telephone advertising and soliciting have aroused resentment against calls from strangers, the aspiring researcher has an immediate handicap to overcome. This is not to say that researchers do not experience skepticism about their motives when other types of contact are attempted, but this hazard is especially acute when using the telephone.

Another limiting aspect to schedules that are completed by telephone is the exclusion of some response forms. Respondents cannot sketch maps, mark positions on lines between two poles, or sort cards by telephone (unless combined with some other techniques; see below). Even using multiple choices is complicated when the response choices are long and/or numerous.

The time factor is particularly critical with a telephone because it is such an interruptive device. Thus, telephone schedules tend to be restricted in length because respondents seldom tolerate long interruptions. The percentage of volunteers who cooperate with any kind of social survey (interview, schedule, questionnaire) is generally inversely related to the length of the time estimated for answering questions. The person who answers the telephone may continue talking for a short time and will reply to a few quick questions, but typically will resent prolonged interruptions.

Another possible draw-back is that telephone schedules normally utilize a sampling frame that depends on telephone directories. This procedure can be a source of complications. One limitation is the fact that not every individual of most defined populations has a telephone (see Chapter 4 for more comments on this complication to sampling). A second, but partially related handicap, is the fact that some people who have a telephone do not have that fact listed publicly. A third potential difficulty is that telephone lists are not organized by location, except by major exchanges. If the researcher needs to include the geographic distribution of individuals into the sampling design, a telephone list, therefore, is unsatisfactory. Although using a telephone schedule to gather data is not necessarily associated with only one type of sampling frame, the common utilization of this sampling design with telephone surveys must be evaluated as a possible limitation.

Using a telephone schedule is a popular technique for collecting data about peoples' beliefs and behavior. It does have many advantages for gathering these kinds of data, but a researcher must regard it nevertheless as a technique to be chosen knowingly and used carefully.

Questionnaires

A questionnaire, which is another method of obtaining data by asking questions, is virtually identical to a schedule of questions. The differences occur in the way the questions are administered and answered. In contrast to oral communication in interviews and schedules, the exchange between the interrogator and respondent in a questionnaire is accomplished in written form. The most common method of delivering questionnaires and receiving them in return is through the mail, although they can be delivered and picked up in person. A mailed questionnaire means the researcher has no direct contact either visibly or audibly with the respondents. These characteristics of a questionnaire give it several advantages and disadvantages compared to schedules.

Advantages

Most of the advantages of a mail questionnaire result from the fact the researcher does not have to go to the place of the respondent. Being able to send the questionnaire through the mail means the respondents are practically as accessible as the nearest mail box. The low cost of communicating by mail, especially when compared to traveling personally to the respondents' locations, gives the questionnaire a significant advantage over interviews and schedules. The fact that the researcher does not administer the questions in person relieves the respondent from having to reply at the time of contact. This may result in more relaxed and carefully considered answers. It also permits using questions and providing for response choices that are longer and more complex than a telephone schedule. Furthermore, the absence of the researcher almost eliminates any apprehensions that some respondents may have about a stranger approaching them either in person or by phone.

Because the informant answers written questions without the researcher being present, a greater objectivity and accuracy in responses can result. All respondents will read the same written words, so there is no likelihood of the researcher clarifying terms differently for different respondents, giving varying emphases to words, or in some other manner affecting the sampled population unequally. Furthermore, many persons are reluctant to discuss taboo topics openly with strangers or to reveal personal facts that can be traced back to them individually. In such situations the anonymity that can be achieved with a mail questionnnaire is a definite advantage.

Disadvantages

Several of the characteristics of mail questionnaires that sometimes make them an effective device for gathering data can, in other circumstances, be a limiting factor. The dependence solely on written words, even in well-stated questions and responses, may actually produce distortions due to the variations in respondents' reading abilities and their subjective interpretations. The results may more than offset the advantages of eliminating the researcher's presence and influence. Although the absence of the researcher may relieve any time pressure on the respondents and allow them to contemplate their answers, it also means that many respondents will never complete the questionnaire (see more on non-response below).

Most of the disadvantages associated with the lack of visual observations, as noted for telephone schedules, are compounded in mail questionnaires by the absence of observations that are audible. Without seeing or hearing the respondent, the researcher is left with the uncertainty about who actually completed a returned questionnaire and under what environmental conditions. There are very few supplementary clues for assessing the reliability of the responses, and no way of verifying the geographic location of the respondent.

Assuming the informant can read, most kinds of response forms can be used. The designer of a questionnaire cannot rely, however, on a strategy of asking questions in a specific sequence to prevent the answer to one question affecting the reply to another. In contrast to schedules, the respondents to a questionnaire can read ahead and respond to questions in any order they wish.

A major difficulty with gathering data through mail questionnaires is their low rate of return and the associated bias in the results. The issue of non-responses is considered here by dealing with (1) bias associated with non-responses, (2) strategies for increasing cooperation, and (3) strategies for adjusting to non-responses.

Bias Associated with Non-Responses

The danger of collecting data from a sample that is biased exists in virtually all field techniques, but the danger is especially critical when the only data are those obtained from questionnnaires that have been mailed back voluntarily. As a result, researchers, amateur and experienced, must be more vigilant to prevent sampling bias in mail questionnaires than in interviews and schedules.

When interviews and schedules are employed, the percentage of completion is normally fairly high. Furthermore, the researcher can usually assess reasons for non-responses. One kind of non-response may result when respondents are unable to reply for one reason or another. These omissions may not generate bias in the data obtained from completed interviews and schedules if they are randomly distributed among the population. A second kind of initial non-response might occur because of the sampling frame. For example, persons may not be home when a field worker arrives for an interview or with a schedule. These people might constitute a distinct subpopulation and as a result their permanent omission would create bias in the data. Because researchers know which individuals are omitted, however, follow-up contacts with this subpopulation can be made. A third kind of non-response may be caused by persons who refuse to cooperate. Non-cooperators often possess distinctive viewpoints. To omit their replies could, therefore, also bias the data. Fortunately, for most populations the percentage of persons who adamantly refuse to cooperate is very small and their impact on the results is minimal. Sometimes the researcher can visually assess some of the characteristics possessed by non-respondents and, hence, partially account for these variables in the analysis.

The situation is far different when a questionnaire is mailed. Some non-responses may result from inaccurate addresses, miscellaneous inabilities of persons to reply, and circumstances that commonly occur randomly throughout the population. If the characteristics of this group of non-respondents is truly random, then it makes little difference in the outcome because the remaining sample can still be representative. The crux of the predicament is that a majority of non-responses occurs because of specific characteristics of individuals—characteristics that often correspond to certain viewpoints being measured by the questionnaire. For instance, persons who fail to respond to a particular questionnaire might be characterized as having attribute X. Attribute X is manifested by non-cooperation with meddling researchers and by holding a polar (say, negative) opinion relative to the topic of the questionnaire. Without these opinions being expressed in the results, the researcher will probably draw incorrect conclusions for the population as a whole. In other words, individuals who do not return their questionnaires are a definite subpopulation that differs from the other subpopulation composed of cooperators. Therefore, the total population consisting of both subgroups is invariably misrepresented by the data based on answers from only one subpopulation. Stated otherwise, when assessing the number of returned questionnaires, the researcher does not need to worry unduly about a low percentage per se but rather about the set of factors associated with the non-response that creates bias.

The researcher using data from mailed questionnaires seldom knows the identity of the non-respondents nor the way they would have answered the questions. Thus, not only is there a likelihood of bias resulting from the non-responses, but the researcher does not have many clues about the characteristics of the subpopulation that failed to respond. Some of the non-respondents may resemble the small group of persons who would not even cooperate with an interview or schedule. More commonly, however, most of them constitute a large group that "just didn't get around to" filling out the questionnaire and sending it back. It is this large group, the specific members of whom are unknown to the researcher, that causes most of the difficulties with bias in mail questionnaires.

A fairly clear demonstration of biased returns is evident from two surveys that asked the same question about marriage partners (Table 6.7). Respondents to one survey (Column A) were obtained from persons who voluntarily mailed replies to Ann Landers. Among this subpopulation,

Table 6.7 Two Surveys with the Same Question having Contrasting Responses
(Reported in the *Lincoln Sunday Journal and Star,* May 29, 1977)

Question		Response (Percentages)		
		A	A'	B
If you had it to do over again, would you marry the same person you're married to now?	Yes	48	70	91
	No	52	30	5
	Don't Know	-	-	4

A - = Survey by Ann Landers
A' = - Survey by Ann Landers, signed returns
B = Survey for *Lincoln Sunday Journal and Star*

composed of persons motivated to send a reply, was a smaller subset (A') who signed their names. Comparing the returns of these two groups exposes the effect of anonymity in answering sensitive questions, namely, that those who willingly signed their replies are more likely to be happily married than those who do not want to reveal their names. Results for the other survey (B) were obtained from a random sample of persons who responded to a telephone schedule in a local area. Comparing the returns from the schedule with those from the mail questionnaire suggests that the subpopulation of persons who felt strongly enough about their marital situation to write a response to the Landers survey differs from the total population, assuming that the localized sample is representative of the same population.[16]

Strategies for Increasing Cooperation

One way of tackling the complications from non-responses is to confront possible non-cooperation as soon as possible. This may be done initially by motivating a reply. When a person is asked at the door or over the phone to answer a few questions, it is often easier for the prospective respondent to comply with the request than not. In contrast, when a person receives a printed request in the mail, it is too easy to discard the materials with the junk mail or to put the letter aside for answering "at a more convenient time". In other words, for many persons it is easier not to reply to a mailed questionnaire than to take the initiative necessary for completing and returning the questionnaire. Therefore, the challenge for the researcher is to attract the recipient's attention and motivate a reply.

One way of motivating a return is to package the request in an attractive manner. The physical appearance of the questionnaire may seem trivial. Consider, however, the tremendous efforts and large amounts of money that are expended by companies to package their products in a manner which will attract customers who glance at them on store shelves. Similiarly, there is evidence that the appearance of a questionnaire can also make a difference in attracting respondents. Various colors, designs, sizes, and introductory announcements have been attempted to attract attention to mail questionnaires.[17] In general, the goal should be a format that attracts the attention of potential respondents and invites their participation in a serious endeavor.

Numerous inducements have been tried to encourage potential informants to return completed questionnaires. At a minimum, the researcher should include an addressed, stamped envelope in which the respondent can return the completed form. Some pollsters have included money and other gifts. Others have tried a form of lottery where laws permit it. Evidence suggests, however, that money spent on telephoning or interviewing sampled non-respondents (see below) is more productive than money included with every mailed questionnnaire. In general, an altruistic appeal is the most effective way of inducing responses. If the potential respondents are convinced that their information and opinions are important and beneficial to society as a whole, they are more likely to return the questionnaire than if they feel it is only gimmickry.

Nevertheless, even after researchers design attractive questionnaires and present strong reasons for respondent cooperation, they usually are disappointed with the low rate of return. What can be done? Normally the first action is to mail a reminder to all non-respondents. This is usually done after a couple of weeks or when the rate of returns dwindles. Being able to send a reminder to non-respondents requires a system of identifying each individual with a particular questionnaire. If responses are kept entirely anonymous, it will be impossible to know who has replied and who has not. This will necessitate sending a reminder to the entire list of potential respondents. Some surveyors have included with each questionnaire a separate identifying postcard which respondents mark and mail at the same time that the completed questionnaire is returned. The identifying postcards, which are not associated individually with the questionnaires, make it possible for the researcher to send reminders to only those persons who have not reported.

A second reminder sent after returns from the first reminder have diminished should probably include a duplicate copy of the questionnaire. If a third reminder is sent, it might be mailed by special delivery. Otherwise, a telephone call, if possible, is a better use of money than additional mailings at special delivery rates.

Even if the percentage of returned questionnaires can be coaxed upwards to a high level (e.g., over 80), this still does not solve the bias associated with non-responses. To repeat, it is not the percentage size of the non-responsive subpopulation that causes complications but rather its distinctive characteristics. Consequently, limited time and resources may be better spent on investigating the characteristics of the non-respondents than on trying to boost the return rate a few percentage points.

Strategies for Adjusting to Non-Responses

One strategy for examining the characteristics of non-respondents is to regard them as a separate population, draw a random sample from this subpopulation, and then administer the questionnaire as a schedule. Using this strategy obviously requires the researcher to maintain a known correspondence between each questionnaire and a particular prospective informant. Correspondence can be achieved by placing a symbol on each questionnaire, but doing this sacrifices the advantages of allowing respondents to remain completely anonymous. Alternately, the researcher can enclose an identification card with each questionnaire and assume that the returned cards correspond to the individuals who have actually returned the questionnaire.

Another strategy, which is somewhat less reliable but cheaper, is for the researcher to keep a close accounting of the date each questionnaire is returned. Results for selected, or all, questions are correlated with (or, regressed on) the return time. If this statistical analysis reveals a trend, then the trend can be extrapolated to the replies that would be expected at the "infinite" time of no returns. This technique has rather severe limitations because its effectiveness depends on (1) finding a relationship between the response types and the slowness of their return and (2) assuming that non-responses are the extreme cases of dilatory returns.

A third strategy for coping with the issue of non-responses to mail questionnaires is to modify the technique by combining the questionnaires with some of the procedures used for schedules. For example, collecting the questionnaires at respondents' homes combines some of the advantages of interviewed schedules with those of the questionnaire and increases the rate of returns over that of voluntary mailings. A second hybrid form of surveying is to mail questionnaires to prospective respondents and then a few days later to phone them for their answers. A third technique is to hand deliver each questionnaire, request that the respondent answer it within a short time while the researcher delivers a few more questionnaires nearby, and then return to collect it. This last technique resembles a schedule, except for the fact that respondents provide information by writing and marking their own answers. None of these entirely solves the complications caused by non-responses, and each modification produces a distinctive combination of advantages and limitations. Nevertheless, the researcher who is willing to attempt new strategies may be rewarded with a high return of reasonably unbiased data.

Evaluation and Applications

Evaluation

Obtaining information by means of a social survey is a data-gathering technique employed for many kinds of research problems. For some kinds of data this may be an inferior technique compared with direct observations because a social survey does not apply to inanimate objects, depends on the cooperation of other persons, filters communication between the researcher and respondent, and assumes that respondents' declarations coincide with their actual behavior or attitudes. In contrast to these limitations, social surveys have the distinct advantage of being able to measure opinions and similar intangible attributes, which cannot be observed visually and must be expressed verbally by individuals possessing these attributes.

After deciding that a social survey is the most appropriate device for gathering data, the researcher must determine the form that is most suitable: an interview, a direct contact schedule, a telephone schedule, or a mailed questionnaire. Criteria for making such an evaluation may vary, but some of the most pertinent factors listed in Table 5.2 are considered here for comparative purposes. Although it is impossible to make precise comparisons because each factor depends on numerous specific details, some contrasts can be suggested for the four forms (Figure 6.1).

1. *Objectivity in wording questions and responses.* When the exact wording of a question is important because a researcher wants to consider wording as a constant factor, the mail questionnaire is best and the interview is weakest. The schedule, in both its direct contact and telephone aspects, normally is fairly objective in the words and text per se but inflections and other linguistic emphases in speaking can produce variations in the way the schedule is administered to different respondents.

2. *Opportunity to clarify meanings.* In contrast to the desire for objectivity in wording, rigidity in wording questions and answers may hamper successful communication because specific words are interpreted differently by various persons. The exchange during an interview provides a researcher with many opportunities to delve into the intended meaning of terms, phrases, and concepts.

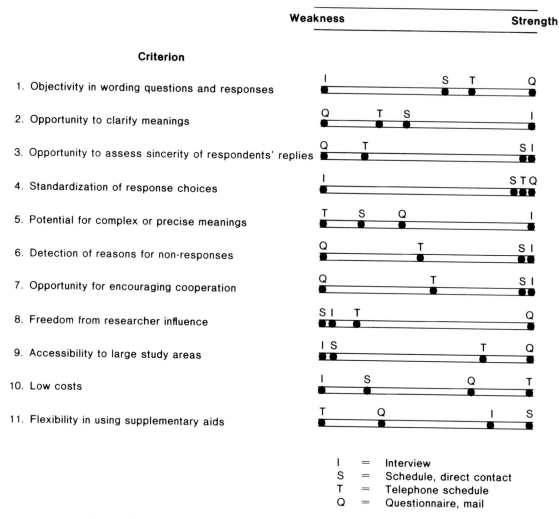

Figure 6.1. Comparisons of techniques for asking questions according to selected criteria

3. *Opportunity to assess sincerity of respondents' replies.* Interpreting non-verbal communication and judging the validity of the information supplied by the respondent is easiest through the direct contact of an interview and schedule. Although a researcher using a telephone can detect a few interpretative clues about the honesty and cooperativeness of the respondent, it is too easy for the researcher to be misled and misinformed by the voice at the other end of the line. There is almost no way of evaluating the validity of responses that come back on a mailed questionnnaire.

4. *Standardization of response choices.* Whenever the informants are asked to reply to a set of choices, the response choices can be standardized to conform with those used in other

research projects. If a researcher wants to standardize all responses, this may be done with schedules and questionnnaires, but almost by definition it is excluded from interviews.

5. *Potential for complex or precise meanings.* The most intricate ideas can be communicated during the conversations of an interview, whereas the predominantly one-way exchange of the direct contact schedule and mail questionnaire may place a restraint on eliciting very complex concepts. Because respondents to a questionnaire can normally take more time to ponder their answers than when a questioner with a schedule is present, the schedule is somewhat more restricted in handling complex ideas. The complexity of questions that can be administered by phone is even more limited because, for most individuals, it is difficult to hear and remember an intricately structured set of questions.

6. *Detection of reasons for non-responses.* The direct contact with potential informants can provide much information about why non-respondents will not cooperate or cannot respond. A verbal exchange on the telephone may supply some information about reasons for non-responses, but this single line of communication limits the number of hints about non-cooperation. Not being able to assess accurately the reasons for non-response by persons receiving mailed questionnaires is a major disadvantage associated with this form of survey.

7. *Opportunity for encouraging cooperation.* When using an interview or schedule, the researcher makes direct contact with prospective informants and can encourage them to cooperate with the research project. The ease with which a telephone contact can be terminated by a prospective informant reduces the researcher's chances for persuasion. Persuasion through the written word sent by mail is the most difficult.

8. *Freedom from researcher influences.* This is almost the inverse of the previous criterion because the potential for the researcher to convince people they should cooperate also means respondents can be affected in other ways. The potential anonymity provided by the mailed questionnaire allows respondents to be extremely frank without feeling threatened in any way. Even though an extremely skilled interviewer may achieve considerable confidentiality, generally respondents are affected by the very presence of an interviewer.

9. *Accessibility to large study areas.* Virtually all places are accessible by mail, provided an address is known. Also, most persons can be contacted by phone in the United States, although the lack of a telephone might be a handicap for certain populations. In contrast, the researcher with a schedule or with plans for interviewing may have difficulty making personal contacts with potential respondents in distant localities and unsafe districts.

10. *Low costs.* The cost of travel and time to conduct personal interviews and schedules often is a major deterrent to the use of these two forms of surveying. Even if travel expenses are minimal, the many hours consumed in interviewing is a sizeable cost factor. In contrast, a telephone survey is usually fast. If the population is local and telephone service is a normal utility expenditure for the researcher, it adds no other expenses. The costs in time and finances for printing, postage, and following up mailed questionnaires can be rather large, but seldom as large as for direct interviewing.

11. *Flexibility in using supplementary aids.* The direct contact occurring in interviews and schedules permits the researcher to utilize maps, pictures, diagrams, and lengthy statements to supplement stated questions. This is not true with telephone schedules. Some supplementary materials may be provided in mailed questionnaires, but bulky items increase costs. More importantly, long responses (e.g., free-story responses to pictures) are very difficult to obtain by mail because most respondents do not enjoy writing lengthy replies.

Applications

The use of interviews, schedules, and questionnaires for collecting geographic data is very common, which means contemporary illustrations of these techniques being applied to research problems are numerous in most geographic journals (e.g., *Annals of the Association of American Geographers, Journal of Geography, Economic Geography, Geographical Review, Urban Geography, Professional Geographer, Antipode, Journal of Cultural Geography,* and *Spatial Analysis*). For personal experience with the application of these techniques, you might do one or more of the following projects:

1. Prepare a short schedule/questionnaire that can be administered directly, by telephone, or through the mails. Strive to include several different response forms.

2. Administer your schedule to a friend three times. Ask your friend to play successively the roles of (a) a cooperative but quiet respondent, (b) a cooperative but very talkative respondent, and (c) a somewhat reluctant respondent.

3. Obtain data from selected informants by using your questionnaire/schedule (a) as a direct contact schedule, (b) as a telephone schedule, and (c) as a mail questionnaire. Keep a complete log/record of the number of persons contacted and the number of completed schedules/questionnaires that are obtained by each technique.

4. Compare the responses you received for specific questions according to the method of administering the schedule/questionnaire. List several possible reasons for any differences that occur.

Notes

1. O.D. Duncan, R.P. Cuzzort, and B. Duncan, *Statistical Geography: Problems in Analyzing Areal Data* (New York: Free Press, 1961), p. 86.
2. Numerous scholars have written about interview techniques; for example, see the bibliography of J.C. Friberg, *Fieldwork Techniques: A Revised Bibliography for the Fieldworker and Reference Guide for Classroom Studies.* Discussion Paper Series, No. 14 (Syracuse, N.Y.: Department of Geography, 1976). References that are particularly helpful include: R.L. Gordon, *Interviewing: Strategy, Techniques, and Tactics,* rev. ed. (Homewood, Ill.: Dorsey Press, 1975); A.C. Kinsey, W.B. Pomerey, and C.E. Martin, *Sexual Behavior in the Human Male* (Philadelphia: W.B. Saunders, 1948), Chapter 2, "Interviewing," pp. 35–62; A. Kornhauser and P.B. Sheatsley, "Questionnaire Construction and the Interview Procedure," in C. Selltiz, L.S. Wrightsman, and S.W. Cook, *Research Methods in Social Relations,* 3rd ed. (New York: Holt, Rinehart and Winston, 1976), pp. 541–573; and M. Parten, *Surveys, Polls, and Samples: Practical Procedures* (New York: Cooper Square, 1966). For an illustration of a short interview, see E.J. Feldman, *A Practical Guide to the Conduct of Field Research in the Social Sciences* (Boulder, Colo.: Westview Press, 1981), pp. 97–112.
3. The term is suggested by A.V.T. Whyte, *Guidelines for Field Studies in Environmental Perception,* MAB Technical Notes 5 (Paris: UNESCO, 1977); but also see Thomas F. Saarinen, "The Use of Projective Techniques in Geographic Research," in W.H. Ittelson, ed., *Environment and Cognition* (New York: Seminar Press, 1973), pp. 29–52.
4. A.L. Delbecq, et al., *Group Techniques for Program Planning: A Guide to Nominal Group and Delphi Processes* (Glenview, Ill.: Scott, Foresman, 1975).
5. L. Mark, "Modeling Through Toy Play: A Methodology for Eliciting Topographical Representations in Children," in W. Mitchell, ed., *EDRA–3* (Los Angeles: School of Architecture and Urban Planning, 1972), pp. 1–3–1 to 1–3–9; F.C. Ladd, "Black Youths View their Environment: Neighborhood Maps," *Environment and Behavior,* 2 (1970), pp. 74–99; K.B. Little, "Cultural Variations in Social Schemata,"

Journal of Personality and Social Psychology, 10 (1968), pp. 1–7; K. Lynch, *The Image of the City* (Cambridge, Mass.: M.I.T. Press, 1960); T. Moore and R.G. Golledge, eds., *Environmental Knowing: Theories, Research, and Methods* (Stroudsbury, Pa.: Dowden, Hutchinson & Ross, 1976).

6. D. Appleyard, "Styles and Methods of Structuring a City," *Environment and Behavior,* 2 (1970), pp. 100–116; A.S. Devlin, "The 'Small Town' Cognitive Map: Adjusting to a New Environment," in G.T. Moore and R.G. Golledge, eds., *op. cit.,* pp. 46–57; T.R. Lee, "Psychology and Living Space," in R.M. Downs and D. Stea, eds., *Image and Environment: Cognitive Mapping and Spatial Behavior* (Chicago: Aldine, 1973), pp. 87–108; R. Maurer and J.C. Baxter, "Image of the Neighborhood and City Among Black-, Anglo-, and Mexican-American Children," *Environment and Behavior,* 4 (1972), pp. 351–388; D.C.D. Pocock, "A Comment on Images Derived from Invitation-to-Map Exercises," *Professional Geographer,* 28 (1976), pp. 161–165; R.A. Sanders and P.W. Porter, "Shape in Revealed Mental Maps," *Annals, Association of American Geographers,* 64 (1974), pp. 258–267.

7. Raymond L. Gorden, *op. cit.,* p. 106.

8. For some illustrations of various strategies, see J. Zeisal, *Inquiry by Design: Tools for Environmental-Behavior Research* (Monterey, Calif.: Brooks/Cole, 1981), pp. 140–154.

9. For an excellent discussion of gaining respondent confidence, see A.C. Kinsey et al., *op. cit.*

10. A.V.T. Whyte, *op. cit.,* p. 65.

11. Prior to this chapter the term questionnaire is used in a generic sense because it is more commonly understood than the word schedule, but in this and subsequent chapters the two are differentiated. Even so, the transitional nature of various forms of asking questions is emphasized by R.L. Gorden, *op. cit.,* who regards them as essentially a single method. Certainly the fact that schedules are transitional to other forms of questioning resembles other data-gathering techniques which have several variations being transitional and difficult to classify neatly.

12. A.R. Blankenship, "The Choice of Words in Poll Questions," *Sociology and Social Research,* 25 (1940), pp. 12–18; R.S. Crutchfield and D.A. Gordon, "Variations in Respondents' Interpretations of an Opinion-Poll Question," *International Journal of Opinion and Attitude Research,* 1 (1947), pp. 1–12; W.E. Deming, "On Errors in Surveys," *American Sociological Review,* 9 (1944), pp. 359–369; R.S. Gorden, *op. cit.; Interviewer's Manual,* rev. ed. (Ann Arbor: University of Michigan, Institute for Social Research, Survey Research Center, 1976); A. Kornhauser and P.B. Sheatsley, *op. cit.;* E.A. Macoby and N. Macoby, *The Interview: A Tool of Social Science,* Handbook of Social Psychology I: Theory and Method (Reading, Mass.: Addison-Wesley, 1954); A.N. Oppenheim, *Questionnaire Design and Attitude Measurement* (New York: Basic Books, 1966); M. Parten, *op. cit.;* S.L. Payne, *The Art of Asking Questions* (Princeton, N.J.: Princeton University Press, 1951); H. Rosen and R.A. Hudson Rosen, "The Validity of 'Undecided' Answers in Questionnaire Responses," *Journal of Applied Psychology,* 39 (1955), pp. 178–181; and K.A. Wang, "Suggested Criteria for Writing Attitude Statements," *Journal of Social Psychology,* 3 (1932), pp. 367–373.

13. N. Erickson, "A Tale of Two Cities: Flood History and the Prophetic History of Rapid City, South Dakota," *Economic Geography,* 51 (1975), pp. 305–320.

14. When only a few cards are assigned to specify in specifically identified classes, the procedure refers to card sorting. When a large number of cards are assigned to the respondent's own set of classes, the technique may be termed Q-sorting because it can feed into a type of analysis known as a Q-technique. For an illustration of card sorting, see E.F. Cataldo, R.M. Johnson, and L.A. Kellstedt, "Card Sorting as a Technique for Survey Interviewing," *Public Opinion Quarterly,* 34 (1970), pp. 205–215.

15. One strategy for selecting a random sample of all telephones (including unlisted ones) in a local area is to use the first three digits, which apply to the specific locality, plus four digits that are obtained randomly.

16. Admittedly, there are other interpretations to these results. Unfortunately, the methodologies of the two surveys were not reported in enough detail to fully evaluate the data-gathering techniques.

17. P.L. Erdos. "How to Get Higher Returns from your Mail Surveys, *Printers' Ink,* No. 258 (1957), pp. 30–31; J.R. Hochstim and D.A. Athanascopoulos, "Personal Follow-Up in a Mail Survey: Its Contribution and Its Cost," *Public Opinion Quarterly,* 34 (1970), pp. 69–81; R.A. Robinson and P. Agisim, "Making Mail Surveys More Reliable," *Journal of Marketing,* 15 (1951), pp. 48–56; and R.F. Sletto, Pretesting of Questionnaires," *American Sociological Review,* 5 (1940), pp. 193–200.

Chapter 7

Sensing with Remote Instruments

This chapter pertains to collecting data by using instruments that can produce visual images of the earth by sensing it from an aerial position. The techniques described in the two previous chapters required the researcher to observe directly the phenomena being measured (i.e., Level 1, fig. 1.4).[1] The emphasis here is on acquiring data which has first been sensed by one or more instruments that can be conceptualized as additional filters between the researcher and the empirical world (Level 2). The primary topic is not about all the many different instruments used to measure geographic phenomena nor on the direct operation of any equipment. Instead, the chapter concerns visible and inherently spatial data that can be obtained from instrumental output.

The output from instruments that remotely sense the earth is normally in a visible form, at least when most amateur researchers view the images. Some special sensors produce data in digital form, but those are not stressed in this chapter. The visible form of the instrumental records does not necessarily mean the phenomena being sensed by the equipment are themselves visible to the naked eye. In fact, a distinct advantage of the technique discussed in this chapter is the extension of observations beyond those that can be made directly by a researcher.

The output of the instruments positioned above the earth consists of images (or other forms) possessing information that is areal. Although photographs are mentioned in other chapters as non-spatial records of visible phenomena, here they are discussed as images from the aerial perspective. The perspective of earth phenomena from a perpendicular position is always areal and hence the images are an excellent source of geographic data. Furthermore, the images provide a synoptic view. A researcher can see the spatial relationships among phenomena in much the same way as when studying maps. This form of output contrasts greatly with the calibrations of most other instruments that measure phenomena by being in physical contact with the objects.

A definite benefit of this locational component, an inherent part of all phenomena observed by remote sensors, is the fact that data are not aggregated by arbitrary areal units. As discussed in Chapter 6, geographers often must accept data aggregated by census districts or similar areal units that lack locational precision. This is not true when the materials are in the form of images from remote sensors.

In summary, this chapter concentrates on the collection of data obtained from the output of just certain instruments, namely, those that produce two-dimensional visible images. The organization of the topics that follow deal with the following general questions:

What are some properties of the electromagnetic spectrum that are sensed by various instruments?

What types of images, that are produced by remote sensors, provide geographic data?

What are some of the variations in the images that result from different carriers of instruments?

How can these images be utilized to solve geographic problems?

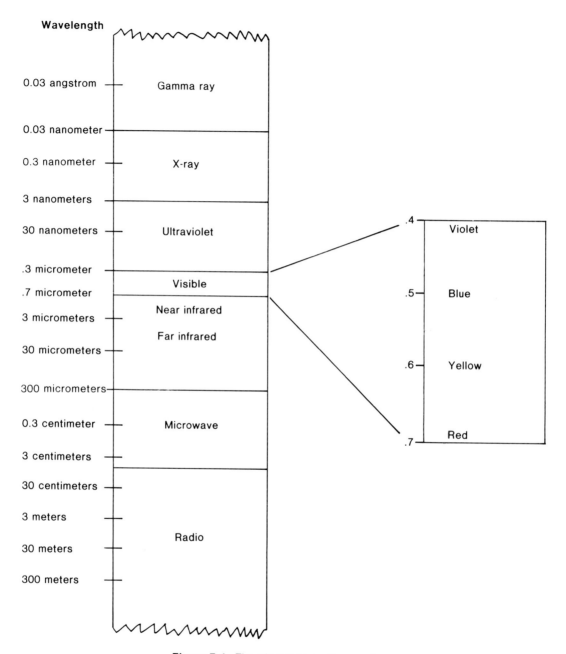

Figure 7.1. The electromagnetic spectrum

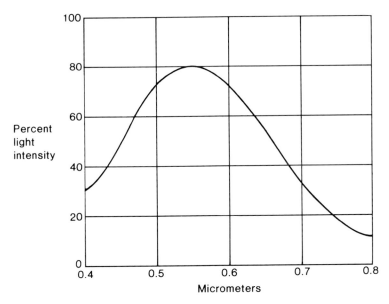

Figure 7.2. Spectrophotometric curve for green

Types of Sensors

Remote sensing refers to the mechanical recording of electromagnetic (EM) radiation reflected or radiated from an object which is not in direct contact with the recorder. A brief summary of the ways this phenomenon (i.e., EM radiation) can be sensed by instruments may be helpful in understanding the kind of data displayed by the images produced by these sensors.[2]

The Electromagnetic Spectrum

All objects (with temperatures above absolute zero) radiate electromagnetic energy which travels in a wave motion. Different forms of electromagnetic radiation are characterized by variations in the length of waves. For example, light, the portion of the EM spectrum which is sensed by our eyes and brain, has wavelengths ranging from 0.4 to 0.75 micrometers (or, microns, i.e., 10⁻⁶ meters; fig. 7.1). Other forms cannot be seen because their wavelengths are shorter (e.g., x-rays and ultraviolet rays) or longer (e.g., microwaves and radio waves) than those visible to humans. Differences in the visible wavelengths are detected as different colors. For instance, if light with a wavelength of approximately 0.55 micrometers strikes the retina, a green color is seen.

Human eyes can detect many shades of color that are all grouped under one color term, for example, green. Although people may be able to compare the colors of two objects by stating that one is more "bluish green" than the other, a more precisely measured comparison is essential for many research problems. This is done by a spectrophotometer, which measures the intensity of light at particular wavelengths and which permits plotting a distribution curve on a graph showing relative light intensity for various wavelengths (fig. 7.2). Each color has its own unique curve differentiating it from other colors and allowing identification by its *spectral signature*.

The same information that produces a spectrophotometric curve also can be used in other forms. For one, the data can be computerized and hence are easily available for additional rapid analysis. Secondly, rather than measuring the entire range of a color curve, only a small range of wavelengths may be selected to produce a type of abridged spectrophotometry. This selectivity is used by some sensing instruments that measure light intensity in only one band, for instance, the green one. The resulting images are restricted to the basic spectral signatures of green. This aids in identifying specific earth features.

The total visible portion of the EM spectrum is very small. In fact, the entire spectrum compares to the visible portion as the equator of the earth compares to the circumference of a pencil. It is not surprising, therefore, that mechanical sensors are used especially to measure parts of the EM spectrum beyond the part detected by the human eye.

The portion of the EM spectrum where wavelengths are slightly longer than those seen by humans (0.75 to about 3 micrometers) is called the *near infrared* band. Although humans cannot see these longer waves, they do affect photographic film differentially. As a result, infrared pictures can be taken and studied in the same way as other aerial photographs (see below).

Beyond the near infrared band is the *thermal* or *far infrared* portion of the EM spectrum (from approximately 3 to 1000 micrometers). These rays of longer wavelengths radiate heat from objects. They vary according to the temperature of a body and its emissivity. As the temperature of an object rises it emits more radiation and in wavelengths that are shorter. In doing so, the spectral curve of an object provides information about its temperature. The amount of radiation depends also on the material itself and its readiness to give up its heat by radiation. Thus, two objects in an identical temperature environment may differ in the amount of heat radiation because of their contrasting composition. This affects emission. By knowing the emissivity of various features (e.g., a concrete walkway is 0.966 and water is 0.993) an interpreter of remote imagery can obtain clues about the objects sensed by the instruments.

Wavelengths even longer than 1 millimeter (1000 micrometers) but shorter than radio waves are called microwaves. They also can provide information on thermal radiation but without some of the atmospheric limitations encountered by infrared rays (see below). These longer wavelengths are used by systems that generate their own energy for sensing the earth, e.g., radar (see below), making it possible to measure earth phenomena without dependence on immediate solar energy. The resolution of most images produced by instruments measuring in this part of the EM spectrum is poor. Therefore, it has limited value for many geographic studies.

Although remote sensors can measure wavelengths in many different parts of the EM spectrum, not all portions are equally useable. The physical separation of the instruments from the ground objects being measured restricts sensing for certain portions of the EM spectrum because the earth's atmosphere absorbs a high percentage of the rays at some wavelengths. Because the rate of absorption differs greatly for various wavelengths, it is necessary to use those portions of the spectrum where transmission is most effective. The wavelengths with high energy transmission through the earth's atmospheric gases are called *atmospheric transmission bands* or *windows*. Thus, remote sensors are restricted to measuring the radiation of earth-based features through those atmospheric windows.

Optical Images

A fundamental aspect of remote sensing is the imagery or physical record of the instruments. A black-and-white photograph taken by a camera from an aerial position is a very commonly used image form because it possesses numerous advantages. An aerial photograph provides a permanent record of all visible phenomena. This allows the researcher to examine the photographic materials whenever desired. The fact that the image can be duplicated permits wide dissemination of multiple copies and consequently wide accessibility to researchers. An air photo can be enlarged to provide different scales. This aids in comparing imagery obtained under varying conditions. Since photographs can be taken rapidly and are relatively cheap, a large number can be acquired to create a wide areal coverage.

Numerous characteristics of photographs can be noted and used to identify earth-based phenomena.[3] Five of these characteristics are tone, shape, size, pattern, and situation. Tone, the distinguishable shades of gray that correspond to different wavelengths radiated from various objects, helps to differentiate phenomena because of their reflection of sunlight. In general, greater reflection appears lighter in tone on positive prints. Another characteristic is shape, which may suggest specific phenomena, as for example, circular fields of crops associated with center-pivot irrigation. In addition to the areal shape viewed from above, the vertical shape may be revealed by shadows, especially if the photograph is taken when the angle of the sun is low. Another geometric property, size, provides clues to the identification of features and yields useful measurements, as for example, the amount of urban land in parking lots. Size is not restricted to just areal dimensions, however, because photogrammetric analysis of objects viewed from a slightly oblique angle provides height information. A fourth characteristic is pattern, which is the repetition in arrangements of features that aids in their identification, especially lines such as those that are associated with roads. Another quite geographic characteristic is situation, which involves observing the combination of identifiable features that are in close proximity to the object under study. This is illustrated by the set of features that distinguish a farmstead from the rest of the rural environment.

The amount of data available from each photograph can be supplemented by simultaneously using several photos. A pair of photographs taken of the same earth area from slightly different positions creates the conditions for stereographic viewing and associated photogrammetric analysis. In a similar manner the combination of a vertically positioned photograph with one taken from an obviously oblique viewpoint allows for greater interpretative power of three-dimensional features. Multiple photos of the same area are especially useful when each records very limited but contrasting wavelengths. The combination of images can be very valuable in the identification of features when the spectral signature of various phenomena are known by an interpreter.

Another way of using several photographs is through a set of airphotos that overlap and thus permit the construction of a composite mosaic covering a large area. A mosaic of photographs provides more information about patterns and situational characteristics than would be available from a single photo at the same scale. Furthermore, comparing airphotos taken of a particular area at different times provides the researcher with dynamic aspects of the visible landscape.

These same advantages are reinforced by color photography, making it easier to differentiate features by their associated shades of color. The use of multiple photographs that sense different bands in the spectrum accentuates features with specific spectral signatures and increases the differentiation of phenomena by various combinations of images. It is through the judicious combinations of images produced by various films and filters that skilled photo-interpreters are remarkably effective in extracting data.

The benefits derived from images sensed by aerial cameras are increased even more when color infrared film (CIR; also called false-color film or camouflage detection film) is used. Objects that appear hazy on regular color photographs because of atmospheric interference may be much sharper on images using CIR film. More importantly, this special film can sense the longer wavelengths that cannot be seen by humans. Because it does, persons can look at images produced by color infrared film and see phenomena not otherwise visible. Because it is impossible to show the "true" color of a wavelength that cannot be seen directly, all colors are altered on false-color film. This alteration is unimportant to a researcher because the primary function of the color images is not to produce a pretty picture but rather to differentiate the reflectance from various objects. On the finished film, infrared becomes red, red appears as green, and green is normally portrayed as blue. With this film, healthy vegetation will appear on the film as red because healthy plants strongly reflect infrared. Unhealthy plants, whether from disease or from being cut for fodder or camouflage, will appear bluish-green. As this illustration implies, color infrared photographs, especially when combined with other pictures of a particular area, provide an interpreter with several opportunities to identify and thus collect data about numerous features.

Other Images

An ordinary camera will not produce a photograph for wavelengths in the thermal or far infrared band, primarily because the temperature of the camera itself radiates some infrared energy that fogs film. Instead, a scanner with a small detector head enclosed in a coolant so cold that virtually no thermal radiation emanates from it senses the EM radiation, which is changed into electral signals of varying intensity. Then the varying current modulates the intensity of a light source which is swept across a strip of recording film or the face of a cathode ray tube (CRT) to produce a visual image. The resulting image shows shades of gray that correspond to various brightness temperatures. These are the actual temperatures combined with the emissivity of objects. These shades of gray can be compared with those produced by sensing some standard thermal sources set at known temperatures, providing a basis for interpreting the brightness temperatures of the observed features.[4]

Another type of information from microwaves uses an active system rather than depending passively on the electromagnetic waves reflected or emitted from the earth. An active system sends out energy for "illuminating" the earth and thus control the wavelengths, direction of transmission, and other characteristics of the energy source. A major kind of active system is radar (*RA*dio *D*etection *A*nd *R*anging). The equipment consists of a transmitter capable of producing waves of the desired length and of sending them out, an antenna to send and receive the energy, a receiver for detecting the return signals, and a display device.

Radar imagery depends on many variables such as the wavelengths transmitted and received back, the ability of objects to reflect radio waves, and the slope and roughness of the land. Because of these difficulties and because of generally poor resolution, radar imagery has limited value as data source for many kinds of phenomena, especially those which can be sensed by other instruments with optical output.

Types of Carriers

Cameras and other sensors can be positioned above the earth's surface by a variety of carriers or platforms. Most individual researchers will not be involved with operating carriers; thus, detailed information about them is unnecessary here, but a brief review may help in comparing the choices of available imagery.

Balloons and Helicopters

Soon after the development of photography (by Daguerre and Niepe in 1839) attempts were made to obtain pictures from aerial positions. Pigeons and kites were not too successful, but balloons soon became useful carriers. Modern balloons are helpful because they can hover over an area, especially if tethered, with no vibration and little movement. They are relatively cheap to manufacture and operate, and there is little risk of damaging instruments in their departure and return to earth. They can carry humans, if manual operators are desired, or they can be equipped with automated equipment for data storage or transmission. Balloons today can operate at heights up to 50 kilometers, a position that offers some of the advantages of spacecraft but without their disadvantages of cost, irretrievability, and restricted movement in an orbital path.

Helicopters, also, can hover over an area, operate at low altitudes, and execute complicated flight paths, all of which make the helicopter an excellent carrier for sensing certain kinds of data. Disadvantages are that they tend to be restricted to sensors for which vibrations are not critical and, like balloons, helicopters are normally restricted to areas smaller than world coverage. In spite of these advantages, neither helicopters or balloons have been used to collect the vast quantity of imagery that is available from other platforms, namely spacecraft and airplanes.

Spacecraft: Rockets and Satellites

Rockets have not been major carriers because of potential damage to instruments during blast off and return impact. Also, the expense of constructing rockets, which photograph for only short periods of time, means the cost per image is very high. Their main contribution to remote sensing, therefore, is in launching satellites and powering spacecraft.

The use of spacecraft for sensing the earth has occurred mostly in recent decades. The first television photograph of earth from space was taken in 1959 (Explorer 6). The first photograph from manned orbital spacecraft was made in 1962 (Mercury 6, 7, 8). These dates reveal that the data collected by instruments carried by these kinds of platforms are relatively new and that techniques for using these data have been developed only recently. In spite of this short period of time, an immense number of images has been acquired.[5]

In July 1972 the National Aeronautics and Space Administration (NASA) put an unmanned satellite in orbit called ERTS–1 (Earth Resources Technology Satellite). It was renamed Landsat–1 just a few days before Landsat–2 was launched in January, 1957. The Landsat satellites travel at an altitude of approximately 918 kilometers in a near-polar orbit from north to south during daylight hours and from south to north at night. They are synchronized with the sun so each orbital pass crosses the equator (north to south) at exactly 9:30 A.M. local (sun) time. By passing over a specific area at exactly the same time the ground conditions of light and shade are similar, although cloud cover and seasonal variations will alter lighting. Landsat takes 103.3 minutes to go around the earth and completes 14 orbits per day by shifting westward 2870 kilometers per pass. However, it is capable of sensing a ribbon of land only 185 kilometers wide per orbit so it takes 18 days to cover the entire earth with a 10–16 percent overlap at the equator (but obviously with much more overlap in polar regions).

Landsat carries two sensors plus facilities for receiving data from special earth-bound collection centers.[6] One sensor is a Return Beam Vidicon (RBV) system which produces black-and-white images. Each image covers an earth area of 185 kilometers square (actually somewhat rhomboidal) with a 10 percent overlap with the adjoining image. The resolution on the few images produced by Landsat 1 and 2 make it possible to distinguish adjacent objects on the ground separated by about 45 meters, but those from Landsat 3 have a nominal resolution of approximately 30 meters.

A Multispectral Scanner (MSS) is the second Landsat sensor. It continuously scans a 185-kilometer swath of earth by making sweeps perpendicular to the direction of travel. The scanner measures radiation like a spectrophotometer in four spectral bands. Band 4 is green (0.5–0.6 micrometers); Band 5 is orange-red (0.6–0.7 micrometers); and Bands 6 and 7 are near infrared (0.7–0.8 and 0.8–1.1 micrometers respectively). This radiation is converted into electric signals and then telemetered to earth, either instantly or after being stored on tape in the satellite for a time. The continuous electrical recordings are separated into sections and converted into images that correspond to ground areas 185 kilometers square. As with the RBV system, MSS images are made on the earth and not on film in the satellite itself.

Another type of carrier is illustrated by the manned satellite, Skylab, which was occupied from May 1973 to February 1974. In addition to some hand cameras operated by the astronauts, it carried a MSS using ten wavelengths in the visible range, two in the near infrared band, and one in the thermal infrared range. Also used were six other photographic cameras; four with black-and-white film using different wavelengths, one with color infrared film, and one with ordinary color film. All photographs were taken from an orbital altitude of 435 kilometers and were aimed at the same area of ground. An active and passive microwave sensor plus a spectrometer were also used to collect data. All of the photographic films and the taped records of non-photographic data were brought back to earth by returning crews. Although manned spacecraft have not been important since 1974, the expected development of permanent satellite platforms with spacecraft shuttling to and from the earth holds great potential for operating manned sensors in the future.

The primary advantages of imagery from satellites are complete world coverage, large areal perspective, and timeliness of the information. The frequent repetition of orbital paths allows continual surveillance of phenomena, providing, for example, data on the status of a crop, both locally and in foreign lands, during a growing season.

Aircraft

Airplanes are the prime carriers of all kinds of remote sensors. They are extremely versatile and possess many of the advantages of the other carriers/platforms. Because airplanes are manned, their sensing equipment does not need to be as expensive as automated instruments and transmitters. Furthermore, the capability of flying at low altitudes, in contrast with spacecraft, permits the use of cheaper instruments with less resolution. Although most of this chapter deals with images produced by remote sensors operated by persons other than the researcher, the use of a light plane for taking large-scale photographs is an exception. It is reasonable for researchers to consider producing their own images from a light plane.

Other aircraft, besides light planes at low altitudes, are flown at altitudes up to 22 kilometers and can remotely sense large areas of the earth and consequently provide worldwide coverage similar to that of orbiting satellites. The imagery resulting from flying over the same area at different heights can be compared. Identification of objects on these large-scale photographs can provide clues for the interpretation of images produced by the more remote sensors.

The relatively low cost of operating standard airplanes means that aerial photographs are accessible to most researchers and are available for most parts of the world. Aviation technology is old enough to have accumulated airphotos for several decades. Thus, with the wide areal coverage and the long period of photographing from planes, researchers are able to choose from a large accumulation of aerial photographs. With the advent of CIR and other imagery from the nonvisible band, the more traditional photographs can be supplemented with a variety of imagery, thereby increasing the amount of information available to interpreters.

Utilization as Field Data

As stated in the introductory section, this chapter concerns the collection of data from images produced by remote sensors, mainly because the images represent phenomena in their field context. The ability to utilize this data source depends not so much on knowing about the physics of the sensing process nor the engineering of the instruments but on the acquisition and interpretation of the imagery. The images per se do not provide answers to geographic problems—but they are an excellent source of data that can be selectively collected, organized, and analyzed to solve research problems. The following section, therefore, is on the availability of remote imagery, the relationship of the images to the features on the earth, and the task of organizing and interpreting observations about the images.

Data Sources

Aerial photographs are such a common form of data that sets of airphotos for local areas are frequently available in libraries, government planning agencies, and private firms involved with photogrammetry. For national coverage the researcher should check with one or more of the many depositories (Table B.2). One of the main centralized sources for aerial photographs and space imagery (as well as maps) is the National Cartographic Information Center. For images from Landsat, Skylab and other manned spacecraft, and aircraft including those flown by NASA, the researcher should contact the EROS Data Center (EDC) in Sioux Falls, South Dakota. Standard order forms, price lists, and forms for a computer search are available from EDC and state NCIC affiliates.

Identification: Determination of Ground Truth

After verifying that aerial photos or other images are available for an area, the researcher must decide whether they portray adequately the phenomena being studied. The basic question is whether the data needed to solve a specific geographic problem can be obtained from images made from remote sensors. The answer depends upon finding a one-to-one correspondence (isomorphism) between features on the earth and variations in images. Realistically the researcher should not expect perfect isomorphism because data errors occur even when observing phenomena directly. Nevertheless, the identification of earth objects by looking at their representations on images should be reasonably accurate. What is accepted as "reasonably accurate" depends on the cost of alternate forms of data collection, the urgency of time, the seriousness of making mistakes, and similar contextual considerations. In any case, the aim is always to reduce the errors of interpretation caused by distortions of mechanical filters intervening between the phenomenon being studied and the researcher.

The core of the technique is to establish the isomorphism between specific ground features and particular image characteristics, and then to extrapolate for all such image repetitions. For example, if five circles of a certain size and hue on a color photograph are known to be exactly where objects X are located, then the interpreter may extrapolate and declare that the remaining fifteen circles of the same size and hue on the image must be other locations of X. Essentially, this technique depends primarily on location and identification.

Establishing a spatial correspondence between earth locations and positions on large-scale aerial photographs is usually not difficult, especially when supplementary locational information is provided with each photograph. In fact, some aerial photographs may serve directly as base maps without additional abstraction making interpretation of location basically a skill of map reading (Chapter 2).

The identification of objects at the specified locations depends on determining *ground truth*. This can be achieved by several methods. One way is to go to the earth location and directly observe the "truth" of what exists on the "ground" (e.g., wheat field, diseased pine trees, school building). The images of these same features can then be regarded as standards for interpreting the remaining part of the imagery set.

Another, but not necessarily mutually exclusive, way is to acquire imagery of part of an area (that is, a sample or test area) using the same sensors but at a much lower altitude than the main set of imagery. Sensing the test area at a closer (lower) position makes it possible to identify certain earth features, and hence determine ground truth. The characteristics of objects on these larger-scale images can then be compared with those on smaller-scale images that cover the entire study area.

A third way is to examine the results from a different sensor or in a different wavelength having at least some features that are more easily identifiable. By comparing relative positions of objects in photos taken in various bands or from multiple sensors, the researcher may make identifications on some images that are more obscure on others.

Published knowledge about the precise location of specific phenomena can aid in determining ground truth. For example, even if Ms. S. has not seen a CIR photograph of Choros before, she can still build a simple interpretive key. By consulting a city map she can correlate the exact positions of paved parking lots and highways with positions on the photograph and then note the image characteristics at those locations.

A way that is somewhat similar and certainly not exclusive of these other ways of determining ground truth is through training by someone who already knows how to interpret imagery. Skilled photo-interpreters are able to detect much information as a result of their previous training and experiences. This knowledge about identifying earth features from image characteristics can be communicated from person to person without having each one re-check ground truth. Ideally a person who anticipates using images should study the techniques of interpretation and gain practical experience while under the supervision of a skilled instructor. This does not mean, however, that the researcher who has not had training in interpreting aerial photographs and other imagery must abandon hopes of using this data source. The amateur who is interested in searching images for a few select phenomena can achieve proficiency in a very short time. A training session of a few hours conducted by a skilled interpreter demonstrating with images having various scales and filters can bring a researcher to a proficiency level quite adequate for identifying selected phenomena. Thus, neophytes should not avoid using the imagery from remote sensors just because they have not had any formal photo-interpretive schooling.

Measurement

The core of data collecting procedures consists of detecting variations in the imagery that indicate the existence of selected phenomena on the earth. When the features being observed are regarded as areally discrete, then the interpretative task may be primarily one of identification. For example, if each patch of ground being irrigated by a center-pivot system is regarded on small-scale images as a point, the researcher's data collecting task usually focuses on identifying all places where these irrigation systems exist. These data are then analyzed for the researcher's particular problems (e.g., the annual increase of irrigators in a region; the density of irrigators in a specified area; the location of irrigators relative to topography, hydrologic regions, or crop patterns).

In addition to being identified on an image, discrete phenomena can be measured according to selected characteristics. If the measurements are to be quantitative, the image must have high resolution and be at a large scale so photogrammetric techniques can be employed.[7] If the measurements are qualitative, then techniques of classification need to be employed (see Chapter 3 and below). When classifying spatially discrete features, the researcher may not require images having as much detail as needed for photogrammetric analysis, but they must be clear enough that the identified objects can be differentiated (e.g., center-pivot irrigation places classified by the crop being irrigated). Likewise, if the phenomena being observed are regarded as areally continuous (e.g., variations that occur internally within an area irrigated by a center-pivot system), the interpretative task is usually classificatory. In some cases there may be subsequent quantitative measurements (e.g., the amount of irrigated crop land that is diseased). In general, the task of classifying areally continuous features is basic to gathering data from images obtained from remote sensors.

The task of scanning images to locate areas that belong to specified classes (e.g., crop types, urban land uses, extent of diseased fields, amount categories of air pollution) can be done by human eye or by mechanical scanners. When humans classify images they may present locational facts in one of two ways: by delimiting distinctive areas, or by classifying each of the minimum areal units that cover the study area. These approaches may be illustrated by an image that consists of three shades (fig. 7.3.a). In most cases the distinction among the various shades will

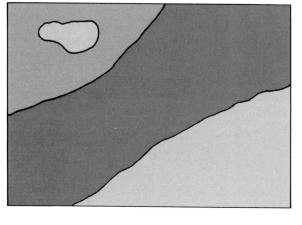

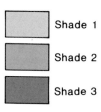

Shade 1

Shade 2

Shade 3

a. Image with three shades of gray

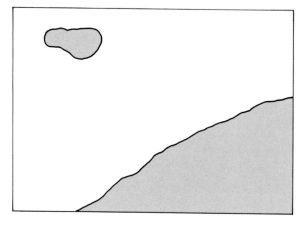

b. Map showing the area of one shade of gray

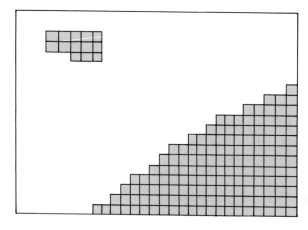

c. Map of pixels matching one shade of gray

Figure 7.3. Classification of an image

not be as obvious as in the illustration, but assume that distinctions do exist, even if only faintly. One way to show the areas where a specific shade (e.g., Shade 3) is located is to trace over or copy the image (fig. 7.3.b). Another way is to divide the entire area into minimum areal units or quadrats or picture elements called *pixels*. Then the researcher decides whether each areal unit does or does not belong to a specified shade class and symbolizes the answer at the corresponding area on a map (fig. 7.3.c). If, rather than using only a dichotomous classification, enough classes are identified to account for all image variations, then the entire map will be covered with various symbols because each of the areal units will be categorized.

These techniques for classifying areas of an image by detecting differences that are visible to the eye are common for interpreting aerial photographs. They may be used also for gathering data from other forms of imagery when the area and/or number of features are limited. For projects that involve classifying a large quantity of imagery, automated and computer equipment may aid in collecting the data.

The input and output for classifying with a computer may be either optical or digital. If data from a remote sensor appear directly as images, that is, as optical outputs, they can be converted to an ordered array of numbers by a digitizer. By scanning each image in a specified sequence of pixels, a detector measures the variations in film density and converts the value for each pixel into corresponding numerals/symbols. The results can be an optical output that resembles Figure 7.3.c. If, instead of having the output in the form of a computer-generated map, the researcher wants the output in digital form, the numbers are suitable for subsequent statistical analysis.

If data from a remote sensor are already digitized (e.g., telemetered data from Landsat), they can be converted to an image or fed directly into a computer for numerical analysis and/or for an areally categorized computer map. Some of the advantages of digitized data are illustrated by those from the MSS in Landsat. The satellite data are transmitted as digital values that may have detected as many as 128 levels of differences, but printing paper is capable of showing only 14 shades of gray. Converting MSS data to images, therefore, loses a lot of detail. Also the MSS scans each 185-kilometer square area with 7,500,000 pixels, producing a total of 31 million data elements for the four bands. If these are transformed into photographs, many details are lost. If the digitized data are utilized directly for classification and analysis, details can be retained.

Computers classify digital data in two ways: unsupervised and supervised classification. Unsupervised classification is theoretically the same as the human categorization by minimum areal units (described above) where each areal unit is assigned to a class, indicated by a numeral. All pixels with similar numericals can then be printed out in map form, showing either the distribution of a single class (like fig. 7.3.c) or the area-filling distribution of all classes. Also, these numerical data can be used like any other digitized computer input for statistical analysis such as calculating the percentage of area in each class or correlating values with other data available for the same study area. The advantages of having the computer classify each pixel on the basis of its image characteristics are clearly in the machine's speed and objectivity. Even though unsupervised classification may be accomplished with electronic speed and mechanical objectivity, it cannot match a human's ability to interpret. The computer may register objectively several pixels as having identical shades of gray, but there is no guarantee that the corresponding earth areas contain identical phenomena. In fact, this illustrates a situation where the researcher is separated from the earth phenomena by so many instrumental filters that the technique hardly qualifies as one dealing with field data.

In supervised classification, the process of classifying is accomplished mechanically by computer, but the researcher selects the features to be matched. This is done by converting the image of the selected feature (say, wheat land) into an electronic equivalency and storing that in computer memory. The computer is then programmed to identify all digital values of the given level as members of the same class. When digitized data are fed into the computer, the computer will match all that are equivalent to the designated feature. The researcher then can interpret the output in terms of the originally specified feature (plots of wheat). If, for example, masses of Landsat data are subjected to supervised classification, the output theoretically could show the world distribution of the specified phenomenon (wheat land).

Although the goals of automated classification are to measure remotely sensed phenomena by classifying the imagery, the level of proficiency often achieved in practical problems has been far less than ideal. In many cases it is necessary for the researcher to do much of the interpretation and classification of imagery directly through visual comparisons by techniques essentially the same as those performed in the field (Chapters 3 and 5). Therefore, the techniques of measuring field phenomena, whether in close contact with them or through remotely sensed images of the phenomena, still depends largely on the observational skills of the researcher.

Evaluation and Applications

This chapter is only a brief introduction to some of the kinds of data available in the imagery produced by remote sensors. It certainly is no substitute for a course in airphoto interpretation or in remote sensing. Nevertheless, researchers should be aware of this pool of data and some of the accompanying advantages and limitations.

Evaluation

The merits of utilizing remote imagery as a general type of data source are summarized here as several advantages and disadvantages. The advantages are (1) its synoptic perspective, (2) its ability to cover large areas in a short period of time, and (3) its extension into the nonvisible, nonaudible realm. The first advantage is that, even though observing phenomena in their areal setting from an aerial position does not require remote sensing instruments, remote sensors invariably provide areal data. This geometric perspective is a tremendous asset to anyone searching for data to solve a problem that is spatial/geographic. Secondly, the availability of data covering large areas, including worldwide coverage by satellites, in a short period of time (even in a few hours by Landsat) generates a tremendous data pool for contemporary world problems. For example, remote sensing techniques have the potential for collecting data that permit a monitoring of the size of the world's wheat crop each year while the wheat is still growing. Lastly, alternative methods of obtaining data about visual phenomena (Chapter 5) and individually expressed human attributes (Chapter 6) depend on the researcher's own senses, but some kinds of data, especially those sensed in the infrared range, require mechanical sensors to transform the attributes into a form that is detectable by human senses.

Disadvantages associated with remote sensing are (1) the dependence on tangible phenomena, (2) the filtering by instruments, and (3) potential difficulty in acquiring specific data. Although one of the advantages listed above referred to the ability of mechanical sensors to collect data beyond the visible wavelengths, they are still restricted to phenomena that emit electromagnetic radiation. As a result, these techniques cannot be used to collect any of the non-tangible data that measure human attitudes, beliefs, and similar attributes. The second disadvantage repeats what was stressed in Chapter 1, namely, that the possibilities for errors may be increased as the researcher is separated from the actual phenomena by mechanical filters. The third disadvantage pertains to the accessibility of desired imagery. On the one hand, many images are restricted by military controls and are inaccessible to the average researcher.[8] Furthermore, the cost of making special flights with sophisticated equipment to collect new data is too expensive for virtually all individual researchers. If a researcher needs data about the location of a particular phenomenon that has not been remotely sensed in a particular band for a specific area at a designated height for a preferred season, then the data are effectively inaccessible. On the other hand, it is not normally the lack of data that causes difficulty but rather the overwhelming abundance of stored data that may make it time-consuming for a researcher to find and select a few appropriate images. Of course, an abundance of data is hardly a major disadvantage—most researchers welcome the luxury of choosing data from a plentiful supply.

Applications

Although aerial photography has been used extensively for only half a century and satellite imagery is less than two decades old, there is no paucity of studies using data from remote sensors. Numerous books, magazines, and research reports are filled with applications of remote sensing.[9] Because aerial photographs and images from other remote sensors provide locational data about many kinds of natural and human features, a list of research problems that could utilize data from remote sensors is almost endless. A few illustrations may typify some of the uses.

A problem in agricultural geography might concern variations in management practices as they relate to farm ownership, distance to market, size of farm, cultural background of farmers, and other independent variables that change from place to place. Management as a research phenomenon is a general term, but it might be operationally expressed in farm fields by such factors as adequacy of moisture, amount of plant nutrients, infestation of diseases, and purity of crops or lack of weeds. By examining color infrared images obtained by a 35 mm camera in an airplane flying at an altitude of 150 to 300 meters, a researcher could probably detect the effects of these management factors on farm lands during the growing season. Plants that are undergoing stress from diseases, lack of moisture, poor fertilization, or lack of moisture would display distinctive characteristics in the CIR images and hence provide a clue to the farmer's practices in managing those lands.[10]

The opportunities for examining large amounts of data is helpful in making inventories of phenomena that exist at certain locations over broad areas of land. Land use maps are available for many cities and other small areas, but these are not universal in coverage and they differ according to classificatory criteria. Imagery from remote sensors that have covered all the land in an almost uniform manner yield data for making maps and inventories of contemporary land uses. These, in turn, can provide a base for developing regional plans. Schemes have been proposed for classifying urban areas using large-scale imagery and general land use/land cover of large regions.[11]

Remote sensors are particularly helpful in monitoring many forms of pollution that are unwanted byproducts of increased population combined with complex technology and industrial production. A growing concern about air, land, and water pollution is expressed by regulations aimed at controlling it, but such control depends on knowing the nature and extent of contamination. Data about contamination are easily acquired from remote sensors because they (1) produce immediate information such as facts about rapidly moving particles in the atmosphere; (2) are capable of covering large areas in a short period of time, for example, about oil slicks, chemical wastes in lakes, and similar materials that disperse over large areas; and (3) can sense some forms of nonvisible pollution, for instance, nuclear radiation.[12]

The vertical perspective is extremely helpful in spotting general patterns and other areal clues to previous settlement features that are difficult to detect by an observer on the ground. For example, evidences of early seventeenth and eighteenth century rice and indigo plantations remain today as various relic features scattered along the coast in southeastern United States. Although these traces can be observed by on-site inspection, it is much easier to gain an understanding of the scattered features by seeing them from the air and observing them as parts of a total pattern. Aerial images permit a researcher to count fields, measure field sizes, and trace former irrigation systems. Also imagery aids in detecting spatial relationships that are less obvious from the more restricted ground view. Images that enhance moisture differences in soils and vegetation reveal that buildings are located on drier areas and agricultural fields on naturally wetter lands.[13]

It is not as easy for the amateur researcher to gain experience independently in this field technique as with some of those presented in other chapters. However, there are opportunities for gaining some experience (1) from workshops and short courses that are held frequently at professional meetings and elsewhere, and (2) from workbooks that are designed to teach skills in interpreting images from remote sensors.[14]

Notes

1. A possible exception is the collection of data by mailing questionnaires to distant respondents rather than administering schedules directly.
2. E. C. Barrett and L. F. Curtis, eds., *Environmental Remote Sensing: Applications and Achievements* (London: Edward Arnold, 1974); R. N. Colwell, "Remote Sensing as an Aid to the Management of Earth Resources," *American Scientist,* 60, No. 2 (1973), pp. 175–183; J. E. Estes and L. W. Senger, *Remote Sensing: Techniques for Environmental Analysis* (Santa Barbara: Hamilton Pub. Co., 1974); D. Harper, *Eye in the Sky: Introduction to Remote Sensing* (Montreal: Multiscience Pub. Ltd., 1976); R. K. Holz, ed., *The Surveillant Science: Remote Sensing of the Environment* (Boston: Houghton Mifflin, 1973); D. Kroeck, *Everyone's Space Handbook* (Arcata, Calif.: Pilot Rock, 1976); T. M. Lillesand, *Fundamentals of Electromagnetic Remote Sensing* (Syracuse, N.Y.: State University of New York, College of Science and Forestry, 1976); J. Lintz, Jr., and D. S. Simonett, *Remote Sensing of Environment* (Reading, Mass.: Addison-Wesley, 1977); P. C. Muerkrcke with J. O. Muehrcke, "Remote Sensing of the Environment, Appendix B" in *Map Use: Reading, Analysis and Interpretation* (Madison, Wis.: JP Publications, 1978), pp. 385–416; R. C. Reeves, ed., *Manual of Remote Sensing,* Vols. 1 and 2 (Falls Church, Va.: American Society of Photogrammetry, 1975); B. F. Richason, Jr., ed., *Introduction to Remote Sensing of the Environment* (Dubuque, Iowa: Kendall/Hunt Publishing Company, 1978); R. D. Rudd, *Remote Sensing: A Better View* (North Scituate, Mass.: Duxbury Press, 1974); F. F. Sabins, Jr., *Remote Sensing: Principles and Interpretation* (San Francisco: W. H. Freeman, 1978).

3. T. E. Avery, *Interpretation of Aerial Photographs,* 3rd ed. (Minneapolis: Burgess Pub. Co., 1978); G. C. Dickinson, *Maps and Air Photographs: Images of the Earth,* 2nd ed. (New York: Wiley, 1979); H. V. B. Kline, Jr., "The Interpretation of Air Photographs," in P. E. James & C. F. Jones, eds., *American Geography: Inventory & Prospect* (Syracuse, N.Y.: Syracuse University Press, 1954, pp. 530–552; L. H. Lattman and R. G. Roy, *Aerial Photographs in Field Geography* (New York: Holt, Rinehart, and Winston, 1965); C. P. Lo, *Geographical Applications of Aerial Photography* (New York: Crane, Russak, 1976); R. M. Minshull, *Human Geography from the Air* (London: Macmillan, 1968); T. J. Smith, ed., *Manual of Color Aerial Photography* (Falls Church, Va.: American Society of Photogrammetry, 1968); and G. B. Sully, *Aerial Photo Interpretation* (Scarborough, Ont.: Bellhaven House, 1969).

4. Detection of radiation having the longer wavelengths of microwaves can be collected also by a radiometer and then directed to a radio receiver. The output is fed through electric currents to create an image, which also displays variations in brightness temperatures. In contrast to the thermal images produced from the far infrared, the images from passive microwaves are more influenced by emissivity of objects than by temperatures. This means that earth features can be differentiated on the basis of their contrasting materials that cause differences in emissivities. For example, ice and icy water radiate very different amounts of energy into a passive microwave sensor. Images from microwaves, therefore, can show the locations of ice floes in frigid seas and areas of open water in mostly ice-packed oceans.

5. The National Archives of the United States has a collection of 2.2 million frames of aerial photographs taken between 1934 and 1942, primarily for U.S. crop surveys. For additional dated but informative comments on available imagery, see D. Kroeck, *op. cit.*

6. Another data-gathering method associated with Landsat satellites (although one beyond the topic of this text) is the Data Collection System (DCS). This system involves numerous automated instruments that are continually measuring characteristics of natural phenomena (e.g., snow, glaciers, soil, lakes) located at generally inaccessible sites around the world. The measurements are made in contact with the phenomena, but the recordings are transmitted to the satellite as it passes over each DCS site and then relayed back to earth when the Landsat is near one of the date receiving centers.

7. American Society of Photogrammetry, *Manual of Photogrammetry,* 3rd ed., Vols. 1 and 2 (Menasha, Wis.: George Banta Co., 1966); and see Note 2.

8. Military uses of remote sensing and the associated ethics of such usage are not considered here, but see S. D. Estep, "Legal and Social Policy Ramifications of Remote Sensing Techniques," in R. K. Holz, ed., *op. cit.,* pp. 362–374.

9. *Air Space; Earth Resources; The Journal of Aerospace Education; Photogrammetric Engineerings and Remote Sensing; The Photogrammetric Record; Remote Sensing of the Environment; Remote Sensing Quarterly.*

10. P. M. Seevers and R. M. Peterson, "Applications of Remote Sensing in Agricultural Analysis," in B. F. Richason, Jr., ed., *op. cit.,* pp. 259–269; and "Crops and Soils," in R. C. Reeves, ed., *op. cit.,* pp. 362–374.

11. J. R. Anderson, E. E. Hardy, and J. T. Roach, *A Land-Use Classification System for Use with Remote Sensor Data,* U.S. Geological Survey Circular 671 (Washington, D.C.: U.S. Government Printing Office, 1972); L. Guernsey and P. W. Mausel, "Application of Remote Sensing in Regional Planning," in B. F. Richason, Jr., ed., *op. cit.,* pp. 319–363; G. K. Higgs, "Application of Remote Sensing in Urban Analysis," in B. F. Richason, Jr., ed., *op. cit.,* pp. 287–318; and "Urban Environments: Inventory and Analysis," in R. C. Reeves, ed., *op. cit.,* pp. 1815–1880.

12. J. E. Estes and B. Golomb, "Monitoring Environmental Pollution," *Journal of Remote Sensing,* 1 (1970), pp. 8–13; and "People: Past and Present," in R. C. Reeves, ed., *op. cit.,* pp. 1999–2060.

13. J. R. O'Malley, "Applications of Remote Sensing in the Analysis of the Rural Cultural Landscape," in B. F. Richason, Jr., ed., *op. cit.,* pp. 239–257.

14. D. Koeck, *op. cit.,;* National Aeronautics and Space Administration, *Earth Photographs from Gemini III, IV, and V; Exploring Space with a Camera; Earth Photographs from Gemini VI through XII; Ecological Surveys from Space; This Island Earth* (Washington, D.C.: U.S. Government Printing Office, 1967–1970); B. F. Richason, Jr., ed., *Laboratory Manual for Introduction to Remote Sensing of the Environment* (Dubuque, Iowa: Kendall/Hunt Publishing Company, 1978); and Carl H. Strandberg, *Aerial Discovery Manual* (New York: Wiley, 1967).

Chapter 8

Using Stored Data

Many geographic field data are obtained by directly observing phenomena (Chapter 5) and/ or asking people about their beliefs and activities (Chapter 6). In addition, original data may be extracted from materials that have been accumulated by other persons and which are now "stored." These materials may include imagery from remote sensors (Chapter 7) or archival collections that contain locational facts. This chapter deals with techniques for collecting new data from this latter source. Subtopics are organized around the following questions:

What kind of "new" information can be extracted from a map?
What data are available through the techniques of content analysis?
What data can be extracted from photographs, tape recordings, and similar non-written
    materials?
Where can a researcher obtain stored materials that may provide locational data?

Selecting Meaningful Data

As a data-gathering technique, the primary purpose for consulting stored materials is to extract "new" information by purposefully reorganizing selected facts from a large complex of data. The selection process is similar to that which occurs whenever a researcher observes phenomena in their environmental context and then measures only chosen features. In the same way, a researcher may commence with documents, maps, audio and video tapes, photographs, and similarly stored materials from which pertinent items are extracted and measured. To acquire data by these techniques involves assembling the materials, assesssing their quality, and establishing procedures for choosing specific facts.

Although many kinds of materials are stored and may provide information about a place, emphasis here is with data that are considered primary. The sources are regarded as primary because they mostly remain in their original recorded form. Although the division between primary and secondary is very transitional, secondary materials are those that have been reworked from their original form. To exclude most secondary materials here is not to imply that they are unimportant to researchers. During the preliminary stages of research projects, secondary materials are very useful because they provide background knowledge about the study area, the research phenomenon, and potential relationships. Furthermore, some of the data sources mentioned in this chapter can be utilized partly as secondary sources irrespective of the techniques used to collect original data.[1] Nevertheless, this chapter deals with stored materials which become essentially a "field" of recorded phenomena from which data are re-gathered.

A major characteristic of data-collection based on stored materials is the intervention of another filter (fig. 1.4). In these circumstances, rather than observing phenomena directly, the researcher relies on the presumably accurate sensing and recording by another person. For example, in contrast to asking a sample of persons to record their daily movements in a travel diary (chapter 9), the researcher may examine the diaries of former travelers that are stored in historical archives. The additional filtering of the second technique occurs because the researcher is seldom privy to information about the writers and their environments. Similarly, photographs taken personally by a field worker will normally provide a better data source than a miscellaneous collection of pictures taken by unknown photographers. Filtering by selectivity and distortion occurs with all techniques, but relying on second-hand data adds the element of uncertainty about procedures used by the person(s) who collected the original data. As a consequence, the researcher must always be very wary of potential flaws in stored materials that are to serve as a source of data.

Criticism is not equivalent to advocating non-usage of this technique. Many phenomena, especially those occurring in the past, cannot be studied otherwise. In fact, virtually all research in the subfield of historical geography is dependent on stored materials, and many research problems can only be solved by resorting to information collected previously.

After the relevant stored materials are found, the researcher should assess them according to the same criteria established for other forms of data collection (Table 5.2). In other words, the researcher should attempt to judge the data collected by another person for degree of validity, reliability, standardization, precision, and completeness. Also, the manner in which locations of individuals are indicated, the minimum areal unit (if continuous phenomena), and the locational precision must be assessed. In many cases it is impossible to evaluate the manner in which the stored data were collected, a situation that generally undercuts the confidence a researcher can place on subsequent use of the data. If these are the only data available, however, then the researcher has no choice but to make certain assumptions about the merits of the data and proceed from that base.

In some cases census figures and similar numerical data with associated locational facts are in a form that is ready for mapping and analysis. More commonly, maps and other materials require reprocessing to produce data that will answer a specific research question. Consequently, a fundamental characteristic of the techniques described here is the selection of specific items from the data source, whether that be a map or other materials.

Selecting "New" Information from a Map

Maps represent the earth positions of numerous phenomena. This abundance of geographic facts permits map readers to extract selectively locational facts about phenomena relevant for just their own problems. As an illustration, the data that are collected by surveyors and map-makers for the main purpose of producing a USGS topographic map can be re-collected to generate "new" information for a different research goal. Field surveyors and cartographers are continually collecting information about the earth position of numerous natural and several cultural features. These include elevation, (i.e., topography), vegetation, ice and water features, political boundaries, transportation routes and facilities, selected buildings (schools and churches), and miscellaneous other features (e.g., cemeteries). The goal of the data-collectors and cartographers is a map that shows the locations of a fairly complete set of features.

Subsequently a researcher may extract from this mass of information a few data that pertain to a specific research problem. These selections constitute a type of creative gathering of geographic information because the researcher often extracts facts that were not directly considered by original fieldworkers. The following examples illustrate the kind of data that can be generated from topographic maps:

the mean distances between towns of various sizes;
the mean distance between rural schoolbuildings;
the density of rural churches and cemeteries;
the network and connectivity indices of highways, secondary roads, railroads, pipelines, and/
    or transmission lines;
the extent of railroad abandonment;
the extent of mining and quarrying;
the extent of canal irrigation;
the size and shape of political units; and
the ethnic background of settlers who first named places.

In addition, these cultural phenomena can be related to each other and/or to natural features in ways such as the following:

the size of towns having specified transportation connectivity;
the relative position of towns and major streams;
the mean distance between rural schoolbuildings in areas of railroad abandonment;
the topographic position of cemeteries;
the size of political units in areas of rough topography; and
the density of rural churches in areas settled by various ethnic groups.

Many of the data listed above involve distance and visual identification/classification, which can be measured without using special equipment. In other words, with a minimum of equipment a researcher can obtain "new" data by carefully extracting locational facts from a previously prepared document/map.

By using more complex instruments, ranging from a simple planimeter to sophisticated digitizers and electronic sensors, the researcher can obtain even more quantitative data. Although this text does not deal with the operation of most instruments, the beginning researcher who wants to extract large amounts of areal data from maps should investigate the feasibility of gaining access to the appropriate equipment.[2]

Content Analysis

The technique of systematically and objectively identifying specified characteristics of a written text is termed *content analysis*.[3] By following a systematic procedure, data which may differ somewhat from the impressions normally obtained by casual reading can be objectively extracted from the contents of written materials. When numerous documents are systematically measured for their contents, data can be aggregated for general information and analyzed for relationships among variables.

A preparatory step in content analysis involves defining the population of materials. For example, a population may consist of all the newspapers published in a certain area for a particular period. Another initial step is to specify the terms that will operationally define the phenomenon

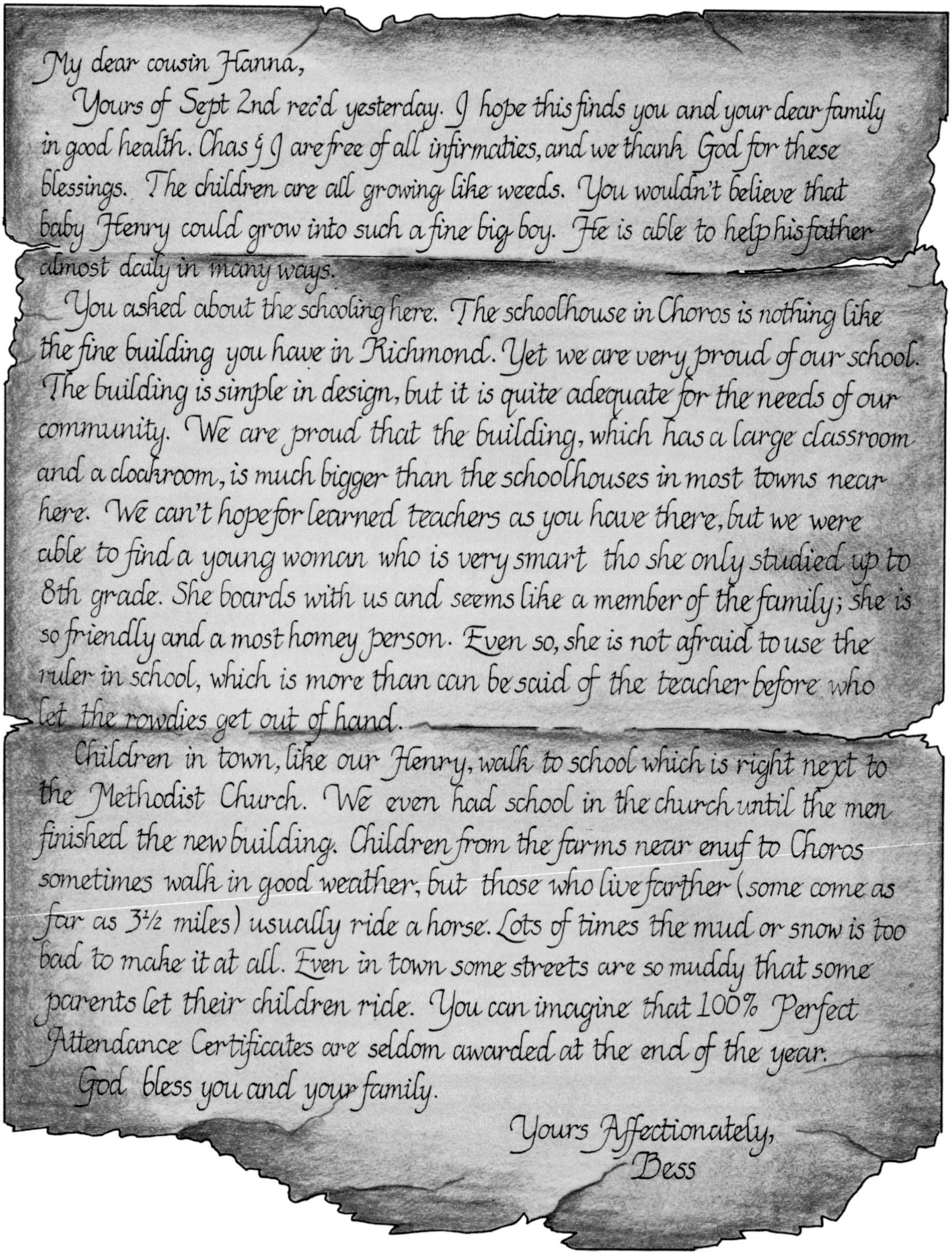

My dear cousin Hanna,

Yours of Sept 2nd rec'd yesterday. I hope this finds you and your dear family in good health. Chas & I are free of all infirmaties, and we thank God for these blessings. The children are all growing like weeds. You wouldn't believe that baby Henry could grow into such a fine big boy. He is able to help his father almost daily in many ways.

You asked about the schooling here. The schoolhouse in Choros is nothing like the fine building you have in Richmond. Yet we are very proud of our school. The building is simple in design, but it is quite adequate for the needs of our community. We are proud that the building, which has a large classroom and a cloakroom, is much bigger than the schoolhouses in most towns near here. We can't hope for learned teachers as you have there, but we were able to find a young woman who is very smart tho she only studied up to 8th grade. She boards with us and seems like a member of the family; she is so friendly and a most homey person. Even so, she is not afraid to use the ruler in school, which is more than can be said of the teacher before who let the rowdies get out of hand.

Children in town, like our Henry, walk to school which is right next to the Methodist Church. We even had school in the church until the men finished the new building. Children from the farms near enuf to Choros sometimes walk in good weather, but those who live farther (some come as far as 3½ miles) usually ride a horse. Lots of times the mud or snow is too bad to make it at all. Even in town some streets are so muddy that some parents let their children ride. You can imagine that 100% Perfect Attendance Certificates are seldom awarded at the end of the year.

God bless you and your family.

Yours Affectionately,
Bess

Figure 8.1. An archival letter analyzed for its contents

Table 8.1　Selected Contents from an Archival Letter

Statements Classified

Favorable

    we are very proud of our school

    it is quite adequate for the needs of community

    we are proud that the building is much bigger than the schoolhouses in most towns near here

    we were able to find a young woman who is very smart

    she is not afraid to use the ruler in school

Neutral

    the building is simple in design

Unfavorable

    the schoolhouse is nothing like the fine building you have in Richmond

    we can't hope for learned teachers as you have there

    tho she only studied up to 8th grade

    the teacher before let the rowdies get out of hand

being studied. For instance, a person might decide that every textual reference to "schooling," "learning," "teaching," "instruction," "education," "pedagogy," and "curriculum," will be regarded as indicators of the general concept of education. Frequently this step involves a concurrent decision about the identification of an individual unit. The unit may be defined variously as a word, a phrase, or a sentence.

Another important step is to specify a procedure for objectively extracting the desired data. This normally requires a set of instructions with enough specificity so that other coders, following these instructions, produce approximately the same results as the researcher. Lastly, the extracted items are measured in some manner such as by counting frequencies of occurrence, by classifying them according to pertinent criteria, or by some other method of measurement (Chapter 3). The results of this last step form the data that can be analyzed with other variables to answer research hypotheses.[4]

Content analysis is illustrated by a project Ms. S. undertook in Choros. She discovered that several years previously the local historical society had requested and obtained private letters from various families who had saved letters from the time when Choros was first settled. She decided to search through that collection of letters to learn about previous attitudes of the community toward educational facilities in Choros. Although this question did not deal with location directly, she believed the historical perspective might aid her in understanding the intensity of current feelings about schooling and the importance of the schoolbuilding issue. Each statement from each letter (for example, fig. 8.1) that mentioned schools, school buildings, school teachers, or schooling in general was copied on a 3x5 card. Then she classified the copied phrases according to a favorable, unfavorable, or neutral attitude toward schools in Choros (Table 8.1). The number of phrases in each category provided information that she could compare with a contemporary survey on attitudes about educational facilities.

In addition to the usual difficulties encountered when classifying transitional individuals into categories, an analyst also faces the uncertainty of not knowing the writers' perspectives. Content analysis requires the researcher to make some assumptions about the context of the writers before valid inferences can be drawn. This is not always easy to do because valid interpretation necessitates being well informed about the settings and conditions of the authors. Ms. S., for example, placed the statement "she is not afraid to use the ruler in school" in the class of favorable statements. Her decision was based on her understanding of the perspectives of schooling in that period of time. If the same statement were classified under contemporary standards, the reference to employing physical punishment would probably be placed in a category other than the favorable one.

The task of reading through all available materials, copying the selected items, and making the appropriate measurements is usually very time-consuming. Like most archival work, this technique is tedious and often without many viable shortcuts. Time can be reduced in some situations through slight modifications. In many cases, a collection may be so voluminous that a researcher will feel a sample of the materials is justified. Also, typing the extracted items directly into a computer terminal for storage and subsequent counting, classifying, and similar organizing may produce a saving in time. When considering this technique for collecting data, the researcher should certainly explore various strategies during the preparatory stages so a time-efficient method for extracting the data is utilized.

Normally the term content analysis refers to the extraction of data from a "field" of primary sources. In a few cases, however, researchers have utilized published results of other scholars in a systematic way to acquire new knowledge. For example, in 1962 Herbert J. Gans reported his study of life in Boston's West End in a book titled *Urban Villagers*.[5] The contents of the publication included much of the data he collected as a participant observer. Later two urban designers extracted information from the book to learn the ways residents of one ethnic group of West End behaved spatially in their rooms, on the streets, and in bars.[6]

Selecting Information from Other Materials

The fundamental techniques of content analysis are also applicable to materials that exist in other than written form. Some variations and modifications are discussed here because they closely resemble content analysis in the way researchers obtain data through specified procedures of extraction.

Selecting Information from Photographs

New insights can be obtained from photographs in the same manner as extracted from maps and written materials. The potential for gathering data from aerial photographs is covered in Chapter 7, but pictures taken from non-aerial positions may also provide locational data. If the position of each scene is known, collections of photographs may contain a wealth of geographic information about visible features. Not only do photographs "store" visual information which can be viewed repeatedly at a later time, but they help expose biased interpretations of visual perceptions. Seeing is a learned process of judging, memorizing, and interpreting visual stimuli, causing humans to "not see" many phenomena in their environments. A mechanically produced photograph bypasses the channeling of visual perceptions and reveals things "not seen" before.

One illustration, although not especially geographic, of how inconspicuous data may be extracted from photographs occurs when political and diplomatic analysts examine a group photograph of the principal leaders in a closed and authoritarian country. Frequently analysts attribute great importance to the arrangement of the persons in the picture because relative positions are a guide to the hierarchical ranking of political leaders.

Another illustration is provided by the nineteenth century photographer who traveled to numerous farmsteads in the Sandhills of Nebraska and took pictures of family groups in front of their sod houses. Now it is possible for a researcher to study those photographs to learn about the environmental setting, including the use of wood resources.[7] Even though it was the photographer who obtained one kind of data originally, the contemporary researcher produced "new" data by selecting previously unused facts from these stored materials.

Selecting Information from Audible Tapes

Another form of non-written material is a magnetic tape that records sound. Tapes were uncommon more than a few decades ago. For this reason their utility as a data source for many historical questions is limited. Nevertheless, even a short time span may reveal trends. For example, analysis of the contents of public speeches that have been preserved by tape can reveal changes in social emphasis, speech patterns, techniques for influencing audiences, and similar data that assist scholars in detecting trends in behavior, including locational activities.

If the definition of "storage" is relaxed, the replaying of conversations taped during data-gathering interviews can be included as a form of content analysis. As mentioned in Chapter 6, an interviewer may choose to tape-record the conversation so that key phrases, emotional terms, inflections and pauses, and other response characteristics that were less apparent during the actual interview can be subsequently analyzed. By repeatedly replaying the tape, the researcher can expand the accessible data by employing some of the techniques that measure characteristics of the communication other than the more obvious contents of the interview.

Selecting Information from Informants

At first glance gathering facts by asking informants to tell about external phenomena appears to be the same as collecting data by interviewing. It is true that techniques for encouraging respondents to provide information are basic to both. Interviewing persons for the purpose of obtaining data about themselves, however, contrasts conceptually with asking respondents to report knowledge they have accumulated about external phenomena. This contrast might be illustrated by two situations. The first one occurs when Ms. S. asks Choros citizens to indicate their personal preferences for the location of future school buildings. In this case each respondent reveals information that only he/she possesses. The filtering is that which exists whenever respondents report contemporary information about themselves to an interviewer.

The second and contrasting situation occurs when Ms. S. asks Dr. A. to tell her the size of the school population. The respondent (Dr. A.) shares a fact that was collected previously which she is now able to recall and relay. In this case Ms. S. attempts to acquire information by tapping the knowledge "stored" in the memory of her informant. The process by which her informant obtained the information, however, underwent all the filtering associated with acquiring and retaining those data in the first place. The additional level of filtering associated with the accumulation of data from informants about phenomena other than their own attributes (Level 3, fig. 1.4) conceptually separates this technique from the interviewing described in Chapter 6.

Collecting data from informants may be advantageous in varying circumstances. One situation arises during the preparatory stages of a project when the researcher sets out to learn about the problem phenomenon, possible explanatory variables, the study area, and background data (Chapter 2). Frequently, while on a reconnaissance, fieldworkers will ask well-informed persons for assistance in acquiring knowledge.

A second situation in which this technique is commonly employed occurs when participant observers depend on key or elite informants to interpret and supplement the data acquired by observations (Chapter 5). Researchers have traditionally depended on a few key individuals who are most willing and able to explain the observable phenomena. This is especially true of anthropologists who spend long periods of time living with a community of persons in a remote region for the purpose of collecting data about various aspects of that society. Dependence on key informants to interpret data collected by observational methods also applies to more commonplace activities. For instance, when Ms. S. made observations of pedestrian pupils, she requested help in interpreting some of the students' actions. By visiting with a few of the adults at school crossings, she was able to gain more insight into the children's paths of movement than if she had depended only on her own restricted observations.

A third situation occurs when researchers who seek information about historical facts rely on persons to recall events and conditions that existed in the past. By asking elderly persons to describe phenomena of a previous generation, researchers may acquire data unavailable from written and other materials. When a large amount of this kind of information is acquired from an informant, the accumulated material is sometimes called an *oral history* which can often be used to supplement and corroborate archival materials of the last few decades.

One form of an oral history focuses primarily on the informant's own activities and beliefs. This makes the resulting record a *life history*, essentially an autobiography communicated orally to a researcher who converts it to written form. Although the best-known biographies based on interviews are those of famous persons, the effective use of this technique for acquiring knowledge about specific research problems usually involves more typical or representative members of a population.[8]

In many respects oral histories, and especially life histories, resemble the general technique of interviewing, especially the free story technique. In this text, description of a life history is separated from other interviewing techniques because additional filtering occurs between the original phenomena reported in a life history and the researcher. Even though the reported events normally involve the informant personally, the tremendous loss of remembered details resulting from the lapse of time makes this kind of data source analogous to archival research. Both involve the happenstance of items that were originally perceived and stored, in memory or on paper, and later recalled or discovered in archives.

Physical Traces and Indirect Clues

Physical traces of human activity remain as a form of stored data that can be selectively observed by researchers. The technique in general resembles others described in this chapter because "new" data are selectively collected through prescribed procedures from phenomena that remain. Procedures for extracting meaningful data differ according to the phenomenon being studied as well as the ingenuity of the researcher.

For some historical data, the procedures for selecting data from physical features are, in a general sense, archeological in nature. The arrangement of bones, sea shells, pottery, charred materials, shaped stones, and numerous other artifacts provide the data for inferring and reconstructing the probable geographies of phenomena in the past.[9] The difficulties associated with acquiring, storing, and interpreting artifacts collected from multiple levels of deposition, which may have been subsequently disturbed by later generations, are complex and necessitate specialized training and experience to overcome. Such specific archeological techniques are excluded here.

Indirect clues to historical events are sometimes available from features that are no longer in their original setting, but nonetheless, preserve certain facts from which geographic implications may be drawn. For example, museum collections may suggest environmental relationships for earlier societies. Previous migrations of people can occasionally be indicated by gravestones.

For contemporary data, the necessity for, yet difficulties in making valid inference may be almost as great as for historical observations. The environmental and temporal conditions in which the physical traces occurred, however, are usually better known. One example of using current traces is the examination of family garbage containers to infer diets and expenditures.[10] Other illustrations are the mapping or recording of paths on campuses, vehicular tracks left in soft-surfaced driveways in rural areas, and scuffed floor wax in art museums for the purpose of measuring aggregated preferences for travel routes and destinations.[11]

Using Census Data

Most amateur researchers are familiar with ways of utilizing figures tabled in various census documents. The "field" in this case conceptually consists of the total set of census facts and the selectivity task only involves finding the proper column of data. In these respects, using census documents is much easier than extracting data from other kinds of sources discussed in this chapter.

The ease of acquisition of census-type data tends to lull users' vigilance against the possibility of faulty data. Too often researchers have accepted published figures without questioning the methods used to collect those data. By adopting facts collected by other persons, many researchers assume that all facets of field work can be by-passed. Viewed in this way, the topic of using census materials is extraneous to this text. On the other hand, every researcher is obligated to evaluate the quality of data being accepted for solving a research problem. From this perspective, using census data is similar to other Level 3 techniques—except that the procedures for selecting facts from the mass of data are mostly unnecessary. The initial collection of data by unknown persons according to unknown procedures pertains to census data just as much as to other techniques involving Level 3 filtering.

The evaluation of census data is not easy because the conditions for their collection are not always reported. For those documents without a section explaining the manner in which data were gathered, the researcher should attempt to estimate the quality of the data. As a guide, a person can apply the questions listed in Table 5.2 to the original field worker. A typical question might be this: "Did the census-taker abide by consistent definitions of all phenomena to which the data apply?" If the researcher is unable to make a reasonable guess about the answers, then this uncertainty should be kept in mind and a note of this doubt should accompany the project report.

Searching for Sources

As the human population increases and societal regulations proliferate, more reports and facts about human activities are collected and stored and more materials are added to the accumulation of documents in the world. Also, with technological developments, information is easier to obtain (e.g., like masses of data acquired by remote sensors). With such masses of data, the tasks of learning about the existence of specific facts and finding their storage places become important. Any researcher who plans to collect data from stored materials must become familiar with common sources and with ways of finding less common sources of primary data.

Learning about the existence of potential data should be undertaken during the reconnaissance phase when other sources of potential data are being examined. There is no unerring procedure that will assure complete success in finding all potential data for a particular research project, but one approach is to begin with organized bibliographical lists. Another is to commence with persons knowledgeable about the specific study area and research problem.

Organized bibliographical lists normally apply to written materials in libraries associated with cities, educational institutions, historical societies, and other special organizations. Specific materials can be located by consulting filing systems of libraries, such as library cards associated with the Library of Congress identification numerals. Due to the distinctive nature of geography, which is concerned with location rather than with a single non-spatial phenomenon, geographic materials under most organizational systems tend to be scattered. This may require browsing in several library locations. Another difficulty is that usually the percentage of primary data may be much smaller than secondary materials, a situation requiring extra search time.

In addition to searching for materials by using library filing systems a researcher should consult one or more bibliographies. Bibliographies exist on nearly every topic. Some are compiled by particular libraries or organizations (e.g., those of the American Geographic Society); some are based on a particular region; still others are topical. Bibliographies vary in terms of language, organization of contents, use of key words, and frequency of updating. Hence, although the "right" bibliography may list all the documents that a researcher needs to examine, finding the bibliography itself may require some searching. This kind of search can be simplified by referring to a bibliography of bibliographies (Appendix, B.3).

The second approach to finding primary sources of data is to request assistance from persons who are knowledgeable about the research problem and setting. This approach may be less direct than using an encompassing bibliography, but it is frequently necessary because many local materials and/or unpublished items are not listed in bibliographies. County records, local newspapers, personal collections of photographs, videotapes of local events, records and reports of churches and social organizations, local and regional maps, and items collected by local historical societies typify the kind of data that are commonly available within a community and not inventoried nationally.

By inquiring from persons connected with local government, historical societies, and similar institutions concerned with the research topic and by following their suggestions about other persons with possible knowledge about materials, a researcher should be able to gain information about many local data sources. In addition to finding potential informants by personal contact, a researcher may also locate knowledgeable persons by placing a notice in local newspapers. This can take the form of a letter-to-the editor, a small advertisement, an entry in the want ads, or even a feature article in a small newspaper. A brief statement about the nature of the study combined with a request for persons having information about the topic to notify the researcher is normally adequate. Sometimes an appeal for old materials will inspire readers to contribute unwanted "junk" that they would have otherwise burned or discarded in the garbage.

Intermediate between local materials and those reported in published bibliographies are miscellaneous reports and documents. These include data collected by state and federal agencies, some public utilities, citizens organizations, and private companies. Researchers who want to re-examine data obtained in past questionnaires, surveys, and polls should consult materials stored with the Inter-University Consortium for Political and Social Research or similar centers.[12] Again, persons at the local level may be helpful in learning about the existence of specific materials, but frequently the researcher must contact the headquarters of specific organizations to obtain copies of materials or permission to examine unpublished records.

Actual acquisition of pertinent materials depends on the form, existence of multiple copies, and the degree of restrictions. Printed library documents may be easy to purchase or duplicate. In contrast, a single but voluminous set of county records may require hand copying under official supervision at the site of storage. In the latter case details about the physical location of specific data items should be noted carefully by the researcher so isolated facts can be rechecked later if necessary.[13]

Evaluation and Applications

Researchers may find that a valuable source of data exists within materials accumulated by other persons. Collecting such data usually involves learning about the existence of stored information, finding its place of storage, assessing its relevance, and developing a procedure for extracting pertinent facts. Although similar steps are basic to other forms of data collection, the process of selecting new information constitutes the core of techniques typified by content analysis.

Evaluation

The advantages and disadvantages of content analysis according to criteria in Table 5.2 will guide this evaluation. The following comments illustrate the application of this technique by Ms. S.

Validity. Communication through a written language, especially when it must bridge the contrasting conditions of different generations, is difficult because words and phrases change in meaning over time. Validity of the researcher's data is diminished if words themselves are misinterpreted. Validity may be sacrificed also by the inferences the researcher places on total contents. Just because many letter-writers in Choros did not mention the school, can Ms. S. correctly infer that they do not feel strongly about education? Furthermore, to establish validity depends on the authenticity of the materials. Scholars and the public have been misled many times by relying too quickly on falsified materials.[14]

Reliability and reproducibility. To produce consistent results from subjectively read materials is a primary goal in content analysis. The degree of reliability varies with the nature of writings and items being extracted, but this technique seldom produces a high level of reliability. For instance, the phrase "it is quite adequate for the needs of our community" might be classified in the neutral category by Ms. S. looking at the same letter (fig. 8.1) at a different time or by another researcher.

Standardization. Few, if any, commonly established standards exist. Researchers have rarely used the same procedures, definitions, and source of data for content analyses as have previous scholars. For example, Ms. S. could not find a previous study that had extracted references to schooling by local letter-writers. She had to develop original definitions and classes.

Precision. Writers use widely varying degrees of precision, so it is difficult to assess the amount of precision in the aggregated results. Certainly the extracted data are no more precise than the least precisely written information. As evidenced by the sample letter examined here, it would be risky for Ms. S. to attempt anything more precise than the three-class measurement.

Detection of errors. Because the process of reading through masses of material is very tedious, researchers seldom recheck their own work. Since there is no other way of catching one's own omissions, misquotations, and similar errors committed in copying selected contents, the technique lacks any built-in safeguard against errors. Errors might be detected, however, if a panel of readers is used for the purpose of achieving greater consistency in all respects.

Locational facts. The lack of locational information is frequently a major disadvantage of this technique because many authors do not write for geographic purposes. Note, for example, that even though Ms. S. knows that the writer of the illustrated letter lived in Choros, she is unable to determine the writer's precise residential location. The writer's reference to the location of the school building as being "right next to the Methodist Church" typifies the degree of spatial imprecision in many records.

Control of variables. Researchers cannot control variables directly. A primary way of obtaining differences is to select writers possessing various characteristics. Unfortunately such facts are often missing in archival materials. For example, if Ms. S. had wanted to correlate educational attitudes with age of father, family religion, and family size, her search for letters containing all these variables would probably have been fruitless.

Control of sample. If there is a large collection of stored materials that are themselves defined as the population, such as newspapers, the researcher can work with a randomly selected subset. Frequently the materials are very limited and are not randomly selected from the entire population. When the population consists of all the documents generated by a public agency, for example, the materials that remain on file for later reference may be very unrepresentative. Similarly, Ms. S. had access only to the attitudes expressed in written form in personal letters saved by families who happened to donate them to the historical society. All persons living in the Choros area during its time of settlement were not potential sources of data for Ms. S. The personalities and attributes of persons who write and/or save letters undoubtedly relate to the phenomenon being studied. This interrelationship causes a bias in the surviving materials that are used as a sample.

Researcher-Subjects relationship. Writers are rarely aware of any research use that may be applied to their writings. As a result, content analysis does not suffer the disadvantages of non-cooperation and affected subjects. This is not to say, however, that writers always express their true feelings in letters, whether private ones or those for public consumption. Just because the technique does not suffer from direct influence of the researcher, it does not follow that stored materials are free of other external pressures.

Size of study area. There is no limitation to the size of area that can be examined through stored materials.

Locational precision. Locational designations usually lack precision (see above).

Sample size. For some studies a small sample can be quite limiting, especially if the available materials are not representative of the entire population. Ms. S. recognized this constraint to her project and wished she had had a wider choice of letters.

Costs. Content analysis is usually very costly in time. For example, Ms. S. had to read through 382 letters to find 30 different families who wrote one or more letters about the Choros schools. Costs of travel, however, are seldom as great as the costs for other techniques which require traveling throughout a study area, unless the pertinent archival materials are located far from the researcher's home site. Ms. S. had to travel to Choros each time she worked with the letters, so traveling distance was like "going to the field" but without traveling within the study area. Costs for purchasing special equipment are rare. If some machine like a microfilm reader is necessary, most libraries possess such equipment. Minor inconveniences might include restricted hours during which libraries, offices, or other storage areas are open for use. In our illustration, Ms. S. encountered very little difficulty, although the society's librarian was a volunteer who was available only part-time and was rather overprotective of the society's possessions.

Recording. The variety of possible forms of recording makes it difficult to evaluate this criterion in general terms. Copying phrases on cards for subsequent classification separates the two decisions of identification and classification, but the execution is tedious. If these two decisions can be made concurrently, the copying step can be omitted and only a tabulation sheet is sufficient. This latter convenience, however, sacrifices any record of the individual data extracted and hence the opportunity to re-examine identification and measurement decisions.

Applications

Several studies that utilized content analysis are cited in the notes. For personal experience with this technique, try one or more of the following problems.

1. Compare the degree of local, regional, national, and international interest of three newspapers by analyzing the contents of their editorials (not syndicated columns) for a specified number of days. Note the frequency and/or proportion of references to events pertaining to these four areal perspectives.

2. Go to the library of a historical society and ask to see one of the library's collection of old photographs. Select a geographic fact that can be obtained from examining a set of such photographs and then try coding 25 pictures for this element.

Notes

1. R. W. Durrenberger, *Geographical Research and Writing* (New York: Thomas Y. Crowell, 1971); L. L. Haring and J. F. Lounsbury, *Introduction to Scientific Research,* 2nd ed. (Dubuque, Iowa: Wm. C. Brown, 1975); T. L. Martinson, *Introduction to Library Research in Geography: An Instruction Manual and Short Bibliography* (Metuchen: Scarecrow Press, 1972); D. C. Pitts, *Using Historical Sources in Anthropology and Sociology* (New York: Holt, Rinehart and Winston, 1972).

2. A. H. Robinson, R. D. Sales, and J. L. Morrison, *Elements of Cartography,* 4th ed. (New York: John Wiley & Sons, 1978), pp. 259–278.

3. R. W. Budd and R. K. Thorp, *An Introduction to Content Analysis* (Iowa City, Iowa: School of Journalism Publications, 1963); O. R. Holsti, *Content Analysis for the Social Sciences and Humanities* (Reading, Mass: Addison-Wesley, 1969); K. Krippendorff, *Content Analysis: An Introduction to its Methodology* (Beverly Hills, Calif.: Sage Publications, 1980); D. Nachmias and C. Nachmias, *Research Methods in the Social Sciences* (New York: St. Martin's Press, 1976); A. V. T. Whyte, *Guidelines for Field Studies in Environmental Perception,* MAB Technical Notes 5 (Paris: UNESCO, 1977).

4. A. J. W. Catchpole, D. W. Moodie, and B. Kaye, "Content Analysis: A Method for the Identification of Dates of First Freezing and First Breaking from Descriptive Accounts," *The Professional Geographer,* 22 (1970), pp. 252–257; H. G. Kariel, "Parochialism Among Canadian Cities," *The Professional Geographer,* 30 (1978), pp. 37–41; D. W. Moodie, "Content Analysis: A Method for Historical Geography," *Area,* 3 (1971), pp. 146–149. Examining the content of diaries is discussed in B. W. Blouet and M. P. Lawson, eds., *Images of the Plains: The Role of Human Nature in Settlement* (Lincoln: University of Nebraska Press, 1975), especially by R. H. Jackson, "Mormon Perception and Settlement of the Great Plains," pp. 137–147; M. P. Lawson, "Toward a Geosophic Climate of the Great American Desert: The Plains Climate of the Forty-Niners," pp. 101–113; and, M. J. Bowden, "Desert Wheat Belt, Plains Corn Belt: Environmental Cognition and Behavior of Settlers in the Plains Margin, 1850–99," pp. 189–201.

5. H. J. Gans, *The Urban Villagers* (New York: Free Press, 1962).

6. B. C. Brolin and J. Zeisel, "Mass Housing: Social Research Design," *Architectural Forum,* 129 (1968), pp. 66–71.

7. In the preface of his book, *Sod Walls: The Story of the Nebraska Sod House* (Broken Bow, Nebraska: Purcells, 1968), R. L. Welsch briefly describes the Solomon Butcher collection of 1500 photographs taken from 1886 to 1892 which he examined eighty years later for evidence of specific features. As a result of his examination of those preserved scenes, he reached "the surprising conclusion" that farmsteads with families living in houses built of sod had a "very high incidence of frame barns and outbuildings" (p. 104).

8. An excellent application of and discussion about this technique is provided by J. M. Freeman, *Untouchable: An Indian Life History* (Stanford, Calif.: Stanford University Press, 1979).

9. F. Hole and R. F. Heizer, *Prehistoric Archeology: A Brief Introduction* (New York: Holt, Rinehart and Winston, 1977); D. H. Thomas, *Predicting the Past: An Introduction to Anthropological Archeology* (New York: Holt, Rinehart and Winston, 1974).

10. R. A. Gould, "The Anthropology of Human Residues," *American Anthropologists,* 80 (1978), pp. 815–835; W. L. Rathje and W. W. Hughes, "The Garbage Project as a Nonreactive Approach" in H. W. Sinaiko and L. A. Broedling, *Perspectives on Attitude Assessment: Surveys and Their Alternatives* (Washington, D.C.: Manpower Research and Advisory Services, Smithsonian Institute, 1975), pp. 151–167; W. L. Rathje, "Modern Material Culture Studies" in Michael B. Schiffer, ed., *Advances in Archaeological Method and Theory* (New York: Academic Press, 1979), pp. 1–37.

11. S. D. Dodge, "Bureau and the Princeton Community," *Annals, Association of American Geographers,* 22 (1932), pp. 159–209; F. H. Forsyth, "The Use of Road Turning in Community Research," *Rural Sociology,* 9 (1944), pp. 384–385; E. J. Webb, D. T. Campbell, R. D. Schwartz, and L. Sechrest, *Unobtrusive Measures: Non-Reactive Research in the Social Sciences* (Chicago: Rand McNally, 1966).

12. The Inter-University Consortium for Political and Social Research is located in Ann Arbor, Michigan. Private companies with similar data are the Louis Harris Political Data Center, Roper Public Opinion Research, and the National Opinion Research Center of Chicago. For other centers, consult the latest edition of the *Research Centers Directory* (Detroit: Gale Research Co.).

13. A study that illustrates the use of Tract Books for information about land claims is by C. B. McIntosh, "Patterns from Land Alienation Maps," *Annals, Association of American Geographers,* 66 (1976), pp. 570–582.

14. P. M. Angle, "The Case of the Man in Love: Forgery, Impure and Simple— 'The Minor Collection: A Criticism'," in R. Winks, ed., *The Historian as Detective: Essays on Evidence* (New York: Harper & Row, 1969), pp. 127–141; A. P. Middleton and D. Adair, "The Case of the Men Who Weren't There: Problems of Local Pride— 'The Mystery of the Horn Papers'," in Robin Winks, ed., *op. cit.,* pp. 142–177.

Chapter 9

Measuring Movement

Geographic questions, which deal with the areal distributions and spatial relationships of phenomena, require data about locations. These data normally are composed of a set of locational facts, each of which specifies the position of an individual at a specified time. Many phenomena, especially people and easily transported objects under human control, change positions frequently so what is a correct locational fact at one moment may not be accurate a short time later. The shifting of locations complicates the task of observing and recording geographic data. This chapter, therefore, deals primarily with techniques for collecting geographic data about phenomena in motion.

Because most techniques for gathering field data in general can also be applied to collecting information about movement, the chapter summarizes many of the techniques discussed in previous chapters. The major treatment here focuses on two basic questions:

What field techniques can be used to obtain data about continuous paths of movement?
What field techniques can be used to obtain data at selected points along paths of movement?

Each of these questions is considered for four general data-gathering methods presented in the preceding four chapters: (1) making direct visual observations, (2) asking questions, (3) employing mechanical sensors and recorders, and (4) using stored materials (Table 9.1).

Table 9.1 Summary of Techniques for Measuring Movement

General Method	Continuous Paths	Points of Movement
Making Direct Visual Observations	Tracking Observations from a fixed position	Observations at origin, destination, or other points
Asking Questions	Travel diary Accompanied walk Recalled responses given in survey	Movement at sampled times Survey at selected points
Employing Mechanical Sensors and Recorders	Film & video-tape Radiotelemetry	Remote imagery Traffic counters
Using Stored Materials	Old diaries Mapped routes	Records for particular places Census materials

Making Direct Visual Observations

The merits of gathering data by visually observing phenomena (Chapter 5) apply also to observing and measuring phenomena as they move from one place to another. Data can be collected about the location of (1) the entire path of movement, in which case the data can be mapped as a continuous line, or (2) selected points, such as the origin and destination, in which case the details of the route are omitted. Whether particular data pertain to a continuous path or to a few discrete points along a route, they result from the researcher's choice of a data-gathering technique rather than the nature of the movement itself.

Although the rate of movement is seldom as important to geographic studies as are the facts of origin and destination and the connecting routes, the aspects of duration and rate do vary tremendously with different phenomena. These different durations and rates are relevant because they may affect the appropriateness of various techniques. For example, the positions of a vehicle on a city street may change several feet per second, but a locational change in a school building may occur only after a duration of several decades. Furthermore, rates of movement may vary within a single journey as illustrated by children who on their way to school move sporadically with bursts of running alternated with temporary halts. Therefore, even though techniques are categorized according to the dichotomy of a continuous path versus discrete points the techniques within each class may vary greatly.

Continuous Paths

One technique for observing and recording information about paths of movement may be termed *tracking*. In many respects it resembles the procedures used in tracking animals, that is, following unobtrusively an individual's movements through an area for a period of time.[1] By observing the person's actual travel the researcher can note the geometry of the path as well as relationships with other phenomena along the route.[2] For studies dealing with pedestrian behavior in various urban environments, the capability of making observations all along the route is important.

A slightly different way of noting continuous paths is by watching from a single position rather than by following after the subject.[3] Often observations made from a position higher than the subject, such as from hovering helicopters, permits an almost perpendicular perspective of the movement within an area. Like tracking, these direct observations provide one of the main sources of data about how people behave spatially as they move individually and collectively through various environments.

Several limitations are associated with observing movement, either from a fixed position or by tracking. One of these restrictions, which also applies to stationary phenomena, is the exclusion of intangible, invisible phenomena. For example, Ms. S., by following school children in Choros, can detect that they tend to move along some streets more than others, but she cannot measure their feelings about the attractiveness of some routes and the avoidance of others.

A second limitation is the normal exclusion of large study areas. There is nothing inherent in this technique that prevents a researcher from tracking a subject across the country or even throughout the world, but such would be impractical—unless maybe one is searching for diamond smugglers. Observations from a fixed position obviously are limited by the distance subjects can be seen. Associated with this areal constraint is the mode of movement. Visual observations from a single position are normally restricted to subjects who do not use motorized power but rather walk, jog, swim, canoe, skate, or cycle. Tracking allows for greater flexibility in mode of travel, but the task of following a specific individual is complicated when the areal range is enlarged beyond walking distances.

A third limitation to this technique is the same that applies to other forms of direct observations; that is, the cost. The amount of time required to follow subjects over routes of travel is normally long enough that it takes many hours to collect data for everyone in a sample set. This constitutes a major investment in securing data. The financial expenses of travel itself may range from the relatively minor costs of shoe-leather to the exorbitant charges for renting a helicopter.

One of the major disadvantages of this technique is the difficulty in achieving an unobtrusive observation. When Ms. S. followed selected children enroute to school, how could she expect to keep them in sight without their being aware that a stranger to Choros was following them? Can any researcher trail a person through a crowd and over a complex route without arousing suspicions? The answer is "yes," but the accompanying difficulties are very great and cause a major handicap in tracking. Obviously, if subjects suspect they are being followed, this knowledge will alter behavior and hence the representativeness of the data. The potential for biasing subjects' behavior (as well as getting the researcher in trouble) are probably greater from tracking than from other forms of observations.

Although the task of remaining unobtrusive while collecting data is not so difficult when observing from a single position, errors tend to increase. Generally, greater concealment of the researcher from subjects causes more visual barriers to collecting accurate data. This is especially true when the goal is to map the paths of individuals who are mixing with other persons as they all move around within a study area.

One difficulty, which may be quite severe when paths are complex and rates of movement are rapid, is the task of recording the data. Because the locational information involves a line—an infinity of point positions—the task of transferring earth positions of the subjects accurately to a map while continuing to watch the traveler's positions requires considerable mapping skill. Also, because several lines on one map may soon become indistinguishable at their intersections, multiple field maps are usually necessary. These in turn require careful organization, handling, and labeling.

A sixth potential limitation is the way movement may affect sampling. A map of stationary objects can be easily prepared and used for selecting a sample of places or objects at specific places. Furthermore, the corresponding objects are still at the mapped location when the researcher goes to the field to gather the data. Such a sampling procedure is not possible, however, when the individuals are in motion. Moving objects cannot be associated unequivocally with a sample area, either on a map or in the field. Zones can be delimited in the field as areas for identifying individuals of a sample, but this strategy frequently produces bias, similar to that associated with administering schedules at selected places (see below).

Points of Movement

If the research goal requires only data from the origin, some intermediate point along the path, or destination point rather than information about the complete route, the task of making direct observations is simpler. One of the most common forms of such data gathering is a *traffic count* that records the total number of pedestrians or vehicles that pass a selected point on a sidewalk or street.[4] Another data set that might be collected is the number and/or characteristics of persons arriving at a store, factory, sports events, public park, or similar destination. If data are collected at the origin of movement, they might concern individuals leaving a city or school building during a specified time period.

When data are collected at points, interest is usually on total numbers (for example, amount of traffic along Highway #10 during periods when children are walking to school). Sometimes facts are needed for the number per category (for instance, number of students in each grade that arrive at the elementary school building by walking). Occasionally the technique is applied to only designated individuals rather than to the aggregate of travelers. When the data pertain to specific individuals, information is often obtained from both places of origin and destination.

Origin-destination studies, which are useful in understanding major flows of persons from place to place, normally deal with aggregate numbers leaving from and arriving at selected areal units or "points" such as cities, or census tracts.[5] In contrast, when direct observation is employed to identify specific individuals at both their places of origin and destination, the researcher may be able to obtain individualized data that provide greater information about patterns of movement. An analogy is the difference between making bird counts at two locations along a migration route as contrasted with the technique of tagging and recapturing birds identified with tags. Although researchers would have difficulty tagging persons, they might convince shoppers to accept identification cards as they enter a delimited study zone and to return the cards as they leave the area.[6]

The disadvantages of this technique can be summarized by weighing the costs of making personal observations against the benefits derived from collecting data directly. For most research problems requiring data obtained at origin or destination points, the facts can usually be collected by cheaper and faster means (see below) than by direct observations. Stated otherwise, the time and effort necessary to make visual observations is normally justified only by collecting the kind of detailed information exemplified by complete paths of movement, not by the more imprecise data of origin-destination studies.

Asking Questions

The important role of social surveys in obtaining geographic data applies equally well to humans in motion. Basically this means obtaining data about movements by asking subjects to report their own movements. The reporting may be accomplished by a variety of forms such as interviews, schedules, and questionnaires, especially when combined with travel diaries or logs.

Continuous Paths

Three techniques for collecting data on paths of movement are (1) recall-type surveys, (2) accompanied walks, and (3) travel diaries. In all cases individuals are asked to report their routes of movement, as well as facts about hypothesized variables, to the researcher. Their statement, whether written, spoken, or mapped, constitute the data. The first technique listed here refers to all interviews, schedules, or questionnaires that elicit information about travel patterns and behavior. Although this technique is applicable for obtaining respondents' expectations about future travel, it is more commonly used to collect data about trips already taken. Consequently, it depends on the recall abilities of respondents.

This dependence on remembering often causes difficulty in obtaining data about past events through social surveys. The severity of forgetfulness depends partly on the nature of the event to be recalled. For example, most persons can remember the place they purchased their last car but they cannot recall where they last drove by a jewelry store. In general, dependence on persons to recall the minute details of their past movements is a data-gathering technique with low reliability.

In spite of this warning, the reader should realize that asking subjects to recall their movements is the best technique for obtaining certain kinds of data. All movements occurring without any other documentation can become a form of data only through the personal recounting of the movement. For example, if Ms. S. wants to investigate where there has been a change in mode of travel to school by elementary pupils in Choros, she will probably need to depend on parental recall. It is extremely unlikely that there are any past records of pupil movements to schools. Ms. S. must rely, therefore, on an adequate number of parents accurately recalling whether their children have walked or ridden to school in recent years.

An *accompanied walk* is one made with a pedestrian, combined with an interview. As the two persons walk along a route, which may be either selected by the subject or designated by the researcher, the interviewer asks the respondent to comment on specified characteristics of the path and/or its environment.[7]

A *travel diary* is a log of movement, which the researcher requests prior to the time of travel and which the subject agrees to maintain for a stated period of time.[8] In a sense the travel diary is a type of social survey because the researcher depends upon another person to verbalize information, which then becomes the data source. Because of this similarity many of the disadvantages of interviews, schedules, and questionnaires also apply to the travel diary. One important distinction, however, concerns the sequence and timing of data collection. For a travel diary the subject knows ahead of time that movement data must be remembered and reported, and the recording is usually done soon after the person has traveled. In contrast, the other forms of social surveys ask the subject after a quite a lapse of time to recall and verbalize events (movements) that may have taken place rather subconsciously.

As a source of data, the travel diary is in some ways superior to direct observations by a researcher. One advantage is the conservation of the researcher's time. Getting subjects to record their own movements saves the researcher a tremendous amount of time in gathering original data. Even though the subsequent organization of materials contained in submitted diaries may take a long time, the initial report of movement is done by the subjects themselves rather than by requiring the researcher to record each bit of information. Also travel diaries allow a lot of flexibility in time. A diary may be kept for any hour, not just when the researcher is able to see the traveler. Diaries can be kept by many individuals for the same time period, whereas a single observer can track only one person (or, at most, a few persons) at a specific time.

Closely associated with flexibility in time is the freedom to use different sampling designs. For direct observations the researcher must sample from persons who are actually in motion at selected times and places. Individuals who happen to be in certain places at particular times are seldom representative of the intended target population. When using travel diaries, however, the researcher has the option of following one of the sampling procedures (see Chapter 4) by which individuals can be selected randomly. There are several different sampling frames that can be employed for selecting individuals who will be asked to keep travel diaries.

A major advantage of a travel diary over using techniques of external observations is the potential for including accompanying unobservable information. Reasons for making a trip, reactions to the environment along the route, and the degree of satisfaction with the journey illustrate the kinds of data that can be obtained by persons keeping diaries but cannot be acquired by merely watching subjects from afar.

In spite of these advantages, the travel diary has limited utility because of severe difficulties in collecting valuable data. A principal disadvantage is the difficulty in gaining the cooperation of selected individuals. The amount of time required to record movements, the tedium of keeping detailed logs, and the inconvenience of bothering with materials to maintain a report are common reasons for non-cooperation. In addition, some subpopulations may not be of help because of age (e.g., the youngest pupils in Choros), language, or cultural barriers.

A slightly different difficulty may be encountered when individuals agree to cooperate but their reported paths are not typical because of modified behavior and/or hesitancy to disclose actual behavior. The modifications of behavior by persons who maintain travel diaries illustrates again a theme that has been stressed for other data-gathering techniques, namely, awareness that one's behavior is being observed and recorded is a potential source of bias. Not only may this awareness change a person's actual movements but it might tempt some subjects to record movements that differ from actuality.

Even when persons cooperate with alacrity and without altering their travel patterns, many find it very difficult to communicate complete and accurate data. The degree of success depends partly on the complexity of information requested, but expecting all persons to describe or map their exact routes of travel is unrealistic. Many persons do not provide adequate information for the researcher to reconstruct and map their routes accurately. A researcher who plans to use this technique must be able to estimate the abilities of respondents and the probable level of accuracy in the data.

Another limitation to this technique of gathering is the possible variability in formats and the resultant difficulty in organizing information for analysis. This organizational handicap may be reduced somewhat by careful planning. A highly structured reporting form designed for a specific research problem not only increases its utility but also tends to diminish reporting gaps and inaccuracies.

Points of Movement

Travel diaries and social surveys can be simplified for data collection when the research question does not require facts about entire routes of travel. Travel diaries can be simplified by requesting locational information only at sampled moments. This can be done by asking each person to carry a Random Alarm Module (RAM), a portable mechanical device that emits a sound after randomized periods of time.[9] The sample consists of a series of facts about location, movement, and whatever else the researcher deems important that the subject records after each beep of the RAM.

This technique is limited by some of the same factors affecting travel diaries such as (1) the difficulty in gaining cooperation from all preferred subjects, (2) the potential bias resulting from awareness of the data-gathering process, and (3) the high probability of some inaccurate and/or incomplete reports. Some disadvantages, in addition to the lack of information about complete paths, relate to the use of RAMs. They involve an expenditure for the equipment, whether rented or purchased, and it is difficult to detect any malfunction in the randomized beep.

Another technique for obtaining travel data is to administer a schedule at selected places, especially origins and destinations. One common place is the home of the person who is asked to report on past and/or expected future movement. Another place is a destination site such as a shopping area, a place of work, and a recreational site. Travelers may be asked questions about the origin, the purpose, the mode, and/or other details about their trips. Another place that is sometimes selected for sampling a population and administering a schedule is an intermediate position between origins and destinations but along a route of interaction. Traffic surveys along highways, streets, and walkways are examples of techniques for collecting data by this technique.

This technique of conducting a schedule at selected places is popular, probably because of the ease of sampling, the potential for generating a large sample in a short time, and the general accuracy of recalled data. Sampling is simplified by this technique because it is easy to define a subset of the population moving along a defined route by selecting every n^{th} individual during a particular period of time. In the case of most traffic flows, this allows a researcher to contact a large number of individuals in a short time. Usually the questions asked of the sampled travelers are easily understood, non-threatening, and pertinent to very recent and easily remembered travel facts. These conditions aid in producing fairly accurate data.

Unfortunately, some of the apparent advantages of this technique can also lead to biased data. Although it is simple to select a sample of travelers along a traffic route during a particular period of time, the choice of time may be quite unrepresentative of the total traffic. Many traffic flows are affected by the time of day, day of week, and season of year. For some characteristics of travelers these temporal variations may not be significantly related to the research objectives. Other attributes, especially the purpose of the trip, often are highly correlated with timing. When this applies, to sample during only one period will produce results that should not be generalized to a total population of potential travelers.

In a similar manner the relative ease of contacting a large number of respondents in a short period of time may lull a researcher into forgetting about the degree of bias that may occur because of non-responses. Often travelers are in a hurry to reach their destinations and do not want to spend time answering questions, so they refuse to cooperate. In terms of achieving a certain sample size, such refusals may not be serious because it is easy to extend the interviewing period and to obtain the responses of additional travelers. In terms of acquiring a sample that is unbiased, however, refusals may be critical because they exclude persons who possess certain characteristics and/or who are undertaking certain kinds of trips.

A different strategy for collecting information about noncontinuous movement is by repeatedly surveying a "panel" of respondents. The most common utilization of a body of informants is to measure changing attitudes, attributes, and activities through time. Professional pollsters sometimes periodically question the same group of persons who have been selected to represent a specified population or community. By repeatedly surveying these same individuals, the pollsters hope to detect trends and other changes through time. If locational data are collected, the same technique can provide changes in positions through time. These changes may range from slow movement as exemplified by place of residence to more rapid changes such as the personal areas of daily activities.

Employing Mechanical Sensors and Recorders

Another way of collecting data about human movements, besides directly observing visual phenomena and requesting verbalized information, is to use mechanical instruments that record movement. Instruments that can sense visual and/or audible stimuli are the primary mechanisms for recording data, but other sensors are possible.

Continuous Paths

Unobtrusive visual observations of movement may be recorded on film or video-tape. Using cameras with film or video-tape to record visual movements is a method comparable to direct observations without some of the disadvantages of requiring the observer to be present at each observation site all the time. The position of the camera can be almost any place such as in a window overlooking a playground of children, on a pole high above a pedestrian mall of shoppers, in a helicopter above city streets, and a variety of other positions that allow for a greater choice of places than normally occupied by a human observer.[10]

Recording movements on film or other permanent sensors has the major advantages of reexamination. As stated earlier in this chapter, the task of mapping numerous paths is quite formidable. Producing a mechanical record of movement does not remove the task of organizing paths of movements for the analytical phase of research, but the capability for replay does allow repeated examination of original motions. It also allows time to produce maps without the pressures and inconveniences often encountered in the field. Furthermore, by replaying the record the researcher can gain objectivity by having other scholars measure and interpret the same data. An additional advantage is that most forms of imagery can be digitized and transferred to computer data by automation. This may mean greater savings in time and more objectivity in measuring characteristics of movement, as well as gaining possible mechanical objectivity.

A disadvantage involves the amount of filtering and distortion caused by the separation of the researcher from the actual occurrence of movement. The severity of this factor depends largely on the role of the instruments in the collection of data. If a researcher watches the movements of children on a playground and films the same activities, then the mechanical record supplements direct observations. If, on the other hand, the researcher sees only the film of playground movements without being able to evaluate associated conditions, then the potential for data distortion is increased greatly.

One conceivable disadvantage of mechanically recorded data is the cost of the recording and replaying instruments. This may not be a critical handicap, however, because the costs of renting equipment that can produce data quickly and reliably may be lower than the value of the researcher's time that would otherwise be expended in collecting and organizing masses of materials over a long period of time.

Another constraint on the technique of filming movement is its general restriction to a small study area. Like a stationary observer, a camera fixed in one location, can collect data from only a limited area. A large study area can be filmed, but only by investing in many cameras.

Discussion here has dealt with films and video-tape because of their ubiquity, but other instruments can aid in gathering data about movements. One is a sensor using thermal infrared film, which can sense the body heat of humans and therefore is not dependent on daylight or visual conditions (Chapter 7). The cost of the equipment is high and the circumstances requiring a non-visual record are rare. For most low-budget research, this technique should probably be dismissed from consideration.[11]

Another technique uses radiotelemetry, an electronic form of tracking. The necessity of planting a transmitter on the subject converts this to a form of spying if the carrier is unaware of being bugged. If the subject agrees to carrying a transmitter for a radio-telemetric record, the ethics of privacy invasion are removed. The subject's agreement, however, introduces potential bias because awareness is often associated with modified behavior. Although valuable data on the behavior of animals has been collected by radiotelemetry, it is rarely employed for gathering data on human movement.

Permitting diarists to verbalize their travel logs audibly into a tape recorder might be considered a type of mechanically recorded data. In this situation the instrument serves only as a convenient way of communicating the informant's message. The instrument does not directly sense movement but only substitutes for a written report. The essence of the technique is still the subjects' reporting of their own movements.

Points of Movement

Two techniques classified in the sixth category (Table 9.1; Row 3, Column 2) are ones using traffic counters and remote sensors. Mechanical traffic counters are routine sources of data. They may count vehicular traffic along streets and highways, subway riders passing turnstiles, and visitors entering tourist centers. It is commonly recognized that these mechanical counters do not measure the exact number of individuals passing them, partly due to a variety of tricks by pranksters. For a study involving a large number of travelers, however, these few errors become rather insignificant. For agencies that need aggregate data on participant movement, these counters are a cheap and adequate data source. For an individual researcher, control over conditions that would channel movement passed one or more counters rarely exists, and the costs of operating such mechanical counters are often prohibitive. If a research problem calls for data of this type, it is usually better to seek such information already collected by others than to attempt original counts.

Acquiring images produced by remote sensors is another way of obtaining data about movement. Much of the imagery available as a data source consists of aerial photographs; thus, the utility of that form is considered here.[12] For example, make the rather unlikely assumption that

Ms. S. has found a set of aerial photographs of Choros taken at nine o'clock in the morning on the fifth day of May for the last eight years. Now ponder on the advantages and disadvantages of these data as indicators of vehicular traffic on Highways #10 and #37 during the time children are going to school.

On the positive side, this is a data source that is very cheap because she is able to examine the stored photographs free of charge. She is able to collect information about total traffic movement in a short period of time and whenever convenient for her. Her facts pertain to highway travel by persons, but she is able to obtain these data without depending on the cooperation of persons making the trips. Her data source is on record, so anyone who wants to challenge or verify her data can do so. The record also permits her to examine past traffic conditions and, as a result, she is not restricted to just the contemporary situation. These are some of the advantages a researcher might expect from a similar collection of images acquired by other remote sensors.

On the negative side, this type of data is limited in several ways. If Ms. S. sought information about the number of passengers per vehicle or some other non-visible phenomenon, she would need alternative sources because she could not get such data from the photographs. Because she can only use what is available, she is restricted to certain kinds of data analysis and research questions. Furthermore, if she considers the available photographs as a sample, she encounters all the problems of selected, non-random data and their associated biases, specifically, the potential biases of generalizing traffic flows from those witnessed on May 5th. As noted previously (Chapter 7), interpretative errors often occur when images are translated into phenomena they depict. For instance, some spots on the highway may be difficult to identify as vehicles. Also, she must assume that each picture is a momentary view of objects in motion, but this assumption may be in error in a few cases. Stationary objects usually appear the same as moving ones in single frame photographs. Whether any or all of these possible disadvantages are severe enough to abandon using this data source is a question Ms. S. must decide within the context of her research objective and alternative sources of data.

Using Stored Materials

This last category of techniques reviews data sources that result from the initial acquisition of information by persons other than the researcher (Chapter 8). In these circumstances the researcher must contend with all the filters (many of which are unknown) inherent in the data-gathering procedure used by the original data gatherer. The variety of data forms is large, but the diversity is represented here by two sources for continuous paths of movement and two for data obtained at discrete points.

Continuous Paths

Maps showing flows of movement along specific routes can be a ready source of data. If the routes of movement are mapped precisely and the number of travelers is shown accurately or given in an accompanying table, there may not be much need for additional manipulation of the data prior to the analytical phase. As with other forms of census-like data, researchers have the responsibility of evaluating the worthiness of available data. This warning is especially applicable for flow maps because depicting movement is cartographically difficult and requires numerous subjective decisions.

Unmapped data about paths of movement are illustrated by informal diaries. The likelihood of finding old travel diaries, kept primarily for research objectives, is extremely unlikely, but the chances of locating personal diaries that contain travel facts are much better. Logs kept by pioneers moving into a new area, personal notes recorded by tourists, and descriptive accounts written by travelers typify this kind of data.[13] Even though records may be incomplete, vague, varying in format, and having unknown authenticity, they may serve as the only source of information about historical movements. In such circumstances, a researcher is faced with the difficult question about whether existing data are adequate for solving a research problem or not.

Points of Movement

A major source of information about migration in the United States is the U.S. Census. Data are tabulated for place of birth, place of residence five years prior to the census year, and parents' birth place. The areal precision of "place" (as well as sample size) varies unfortunately from census to census. For migration studies not requiring a high level of locational precision, census data are a primary source.[14]

Origin or destination data, which can provide historic data on human movements, may be collected and stored in many places. Some of these sources are the following:

the home address of each worker in a factory, business firm, public institution, and government agency;

the home address of visitors to tourist and entertainment sites;

changes in residential addresses listed in farm and city directories;

changes in the residential addresses of families attending school, having utility connections, registering property with a government agency, and listed by credit companies; and

the registration of students in classes located in specific classrooms within a campus area.

The great variety in ways these data are obtained means it is difficult to generalize their merits. Researchers must attempt to evaluate the specific conditions associated with a particular data set deemed potentially useful. In some cases the keepers of the data file may be unable, either because their responsibility is quite distinct from the data collectors or because of confidentiality regulations, to provide information about the procedures followed in acquiring the data. In fact, many data sets are entirely unavailable to outsiders because facts are regarded as confidential information.

Evaluation and Applications

Evaluation

Collecting data about the phenomena in motion is normally more difficult than gathering facts about phenomena not in motion. Otherwise, the comparative merits of the various techniques are similar to those discussed in previous chapters (Chapters 5–8). Nevertheless, each technique contemplated as a method for gathering data should be evaluated by using Table 5.2 before accepting a particular one for use in a research project.

Applications

In addition to reading about the application of specific methods by other scholars, you may wish to gain practice in some of these field techniques by doing one or more of the following problems.

1. Map the complete movement of a player in an athletic game for a short period of time. Be sure to have a plentiful supply of maps so the paths of movement can be shown clearly. During another short period, show the locations of the same player at each ten-second interval. Discuss the relative merits of these two techniques for observing and recording data about movements.

2. For a specified period of time (e.g., 20–30 minutes) count the number of pedestrians moving past a position on a sidewalk located in the center of a city. Repeat this procedure for a few other positions in the same general vicinity. Discuss some reasons for the differences in the counts at the various positions. Include in your comments information about sampling bias and other factors of gathering the data that may cause different pedestrian counts, as well as reasons for actual differences in pedestrian flows.

3. Keep a complete travel diary of your own outdoor movements for a 48-hour period. Record the time, origin, route, destination, and mode of travel. Obtain a similar travel diary from a friend. Discuss the effectiveness of this technique in obtaining data about the parts of movement by persons.

Notes

1. G. C. Sanderson, "The Study of Mammal Movements—A Review," *Journal of Wildlife Management,* 30 (1966), pp. 215–253.
2. L. A. Hoel, "Pedestrial Travel Rates in Central Business Districts," *Traffic Engineering,* 38 (1968), pp. 10–13; G. H. Winkel and R. Sasanoff, "An Approach to an Objective Analysis of Behavior in Architectural Space," in H. M. Proshansky, W. H. Ittelson, and L. G. Rivlin, eds., *Environmental Psychology* (New York: Holt, Rinehart and Winston, 1970), pp. 619–631.
3. Dietrich Garbrecht, "Pedestrian Paths Through a Uniform Environment," *Town Planning Review,* 42 (1971), pp. 71–84. For bibliographical references on pedestrian behavior, see K. S. Bartholomew, *Pedestrian Movement: Selected References 1965-June 1972* (Evanston: Transportation Center Library, Northwestern University, 1972); D. Garbrech, *Pedestrian Movements: A Bibliography,* Exchange Bibliography No. 225 (Monticello, Ill.: Council of Planning Librarians, 1971); M. R. Hill, *Pedestrian Behavior and Facilities Design: Selected Bibliography,* 1970–1975, Exchange Bibliography No. 1112 (Monticello, Ill.: Council of Planning Librarians, 1976).
4. D. Emmons, *The Pedestrian Count* (Chicago: American Society of Planning Officials, 1965); D. R. Drew, *Traffic Flow Theory and Control* (New York: McGraw-Hill, 1968), see especially pp. 85–89.
5. E. J. Taaffe and H. L. Gauthier, *Geography of Transportation* (Englewood Cliffs, N.J.: Prentice-Hall, 1973.)
6. G. R. Woods, "A Pedestrian Origin-Destination Survey for the Siting of a Pelican Crossing," *Traffic Engineering and Control,* 16 (1975), p. 328.
7. K. Lynch and M. Rivkin, "A Walk Around the Block," *Landscape,* 8 (1959), pp. 24–34.
8. A familiar illustration of the use of non-travel diaries is the collection of data about the choice of television programs. The utilization of diaries on travel is explained in W. L. Garrison, et al., *Studies of Highway Development and Geographic Change* (Seattle: University of Washington Press, 1959), pp. 168–170.

9. M. W. Martin, Jr., "The Use of Random Alarm Devices in Studying Scientists' Reading Behavior," *I.R.E. Transactions on Engineering Management,* EM-9 (1962), pp. 66–71; M. W. Martin and R. L. Ackoff, "The Dissemination and Use of Recorded Scientific Information," *Management Science,* 9 (1963), pp. 322–335.

10. R. L. Carstens and S. L. Ring, "Pedestrian Capacities of Shelter Entrances," *Traffic Engineering,* 44 (1970), pp. 38–43; J. J. Fruin, "Designing for Pedestrians: A Level of Service Concept," *Highway Research Record,* No. 355 (1971), pp. 1–15; K. Lautso and P. Murole, "A Study of Pedestrian Traffic in Helsinki: Methods and Results," *Traffic Engineering and Control,* 15 (1974), pp. 446–449.

11. Another instrumental technique of limited value for geographers is the measurement of paths across the floor of a room by an electronic system called a hodometer that records the location of all footsteps; see R. B. Bechtel, "Human Movement and Architecture," in H. M. Proschansky, W. H. Ittelson, and L. G. Rivlin, *op. cit.,* pp. 642–645.

12. Note that aerial photography is not uniquely associated with "instrumental observations of points along paths of movement." If the researcher personally takes the pictures, they are only adjuncts to "direct observations". If, in another case, the filming is done continuously, the photographic data are not restricted to discrete places.

13. M. P. Lawson, "Toward a Geosophic Climate of the Great American Desert: The Plains Climate of the Forty-Niners," in B. W. Blouet and M. P. Lawson, eds., *Images of the Plains: Role of Human Nature in Settlement* (Lincoln: University of Nebraska Press, 1975), pp. 101–113.

14. R. D. Larkin and G. P. Peters, *Population Geography, Problems, Concepts, and Prospects* (Dubuque, Iowa: Kendall/Hunt Publishing Company, 1979); E. Moore, *Residential Mobility in the City,* Commission on College Geography, Resource Paper No. 13 (Washington, D.C.: Association of American Geographers, 1972), especially "Appendix: Sources of Data for Mobility Studies," pp. 47–48; C. C. Roseman, *Changing Migration Patterns Within the United States,* Commission on College Geography, Resource Paper No. 77–2 (Washington, D.C.: Association of American Geographers, 1977); H. S. Shryock and J. S. Siegel, cond. ed. by E. G. Stockwell, *The Methods and Materials of Demography* (New York: Academic Press, 1976), especially Chapter 21, "Internal Migration and Short-Distance Mobility," pp. 373–405.

Chapter 10

Developing a Research Plan

Collecting data is only one of the several stages in solving a research problem. The relationship of this stage to the total research cycle is introduced in Chapter 1 (fig. 1.3). Comments about the interdependence of field techniques with other parts of a research project permeate the entire text. In this final chapter the relationships of data collection to the total research project is reviewed, especially as it is expressed in a research plan.

This chapter, which pertains to the preparation of a research plan, concentrates primarily on the following questions:

What is the general role of a research plan?
What are the elements in a basic research plan?

Role of a Research Plan

A research plan for a particular project is the comprehensive guide that specifies the general method as well as the detailed procedures for solving a research problem. The plan is essentially a working paper that assists the researcher in progressively clarifying and integrating the detailed definitions and procedures with the overall research objective.

An Expression of the Research Objective

The overall research objective might be considered as being analogous to a tree. The philosophical base is the unseen root system; the methodological approach is the major trunk; and the detailed procedures are the branches. The bulk of a research plan normally concerns the many "branches," but these are simply the more obvious expressions of a supporting methodological core and a philosophical root. Likewise, even though most planning time is consumed by making numerous preparatory decisions for a research project, underlying assumptions should not be neglected.

The philosophical root for most research is not stated because empiricism (or positivism) dominates most research. In fact, techniques presented in this text, with the possible exception of experiential fieldwork, are ones that are most appropriate for gaining knowledge through the scientific and empirical methods of the positivist philosophy. Persons whose major objective is solving societal problems are interested mainly in choosing techniques that will be most effective in collecting and analyzing data that provide meaningful answers. For them, questions about ways of knowing and similar philosophical topics are regarded as minor when compared to the pragmatic task of solving a stated problem. Other persons, however, recognize that even the manner in which a problem is stated implies certain underlying assumptions about knowledge. Although this text does not deal with various relevant philosophies, careful scholars are cognizant of the implicit roots of their research.[1]

The methodological core of research is usually more evident than its philosophical base. Most researchers realize that a variety of procedural questions are outgrowths of a core research objective. Many decisions that must be made during the preparatory stages of a research project are highly contingent on the overall objective of the undertaking.

Objectives can be classified by various criteria such as personal value system, methodological system, or paradigms of a particular discipline. In this text, however, four general types of objectives are presently only as rough categories that differentiate some approaches to the study of geography (Table 1.1). The four types are described in terms that may help non-geographers understand something about the diverse styles of geographic studies. A majority of contemporary research projects requiring empirical data are those that strive to develop principles of location (Type II) and apply geographic knowledge to specific situations (Type III).

These methodological approaches (as well as the philosophical assumptions) are expressed, either explicitly or implicitly, in the research plan. The relationship between the approach and the plan can be illustrated by referring to the school building issue in Choros.

The Choros school board expects to locate the building(s) according to a variety of factors such as availability of land, land values, areas of noise and traffic congestion, future patterns of city growth, and the safety of pupils who walk to school. The last factor is selected here to demonstrate how the application of principles may generate the need for certain data. First the school board accepted the following objective:

> The goal of the board is to minimize the probability that a pedestrian pupil enroute to and from school will be injured by a motorized vehicle.

To meet that objective the board accepted these principles or generalizations:

  a. There is a greater probability that an elementary age pupil will be injured by a motorized vehicle when that child must walk across a highway.
  b. As the number of pupils crossing a highway increases, the likelihood of one or more being injured is increased.
  c. The location of the elementary school(s) will determine the number of pupils who must cross a highway when walking to school.
  d. One of the ways of reducing the likelihood of pupils being injured is to locate the school(s) where the least number of pupils must cross a highway.

Therefore, the best location, according to this particular set of principles, depends on the residential locations of Choros pupils. The data that are needed are the home sites of the Choros students. The precision of these locational data must be given at least in terms of their home relative to the two intersecting highways.

Ms. S. became interested in the locational problem in Choros from a different approach. She wanted to study it for the purpose of learning more about where school buildings are located relative to such factors as land values, areas of noise and traffic congestion, future patterns of city growth, and accessibility to pupils. When she considered this last factor, for example, she initially hypothesized that the citizens would ultimately choose a position close to the point of minimum aggregate travel for pupils. After doing some preliminary checking during her field reconnaissance, she decided to modify her investigation of this subtopic by concentrating on the present nodal region of pedestrian pupils. Her revised hypotheses were typified by the following.

a. The pedestrian routes of pupils do not coincide with the most direct paths between residences and school.
b. The number of pupils using pedestrian routes crossing a highway varies by location.
c. The pedestrian routes of pupils that deviate the farthest from direct-line paths between residences and schools are those associated with the routes having the greatest number of pedestrians where they cross a highway.

Irrespective of your judgment about the plausibility of her hypothesized relationships, you can see that once Ms. S. formulated some hypotheses her data needs were set. In other words, the statements she proposed as possible relationships among phenomena generally determined the nature of the empirical facts needed to support or reject the generalizations.

The degree to which researchers actually specify the methodological bases for their research varies considerably. Many who apply geographic principles to a situation like Choros do not formally state them as listed above. Likewise, some forms of research that fit the Type II approach are not necessarily amenable to stating formal hypotheses. In spite of this diversity, however, a research plan that specifies the ways a problem is to be solved expresses an implicit overall research objective.

Timing for the Plan

Although important, the methodological background does not form what is normally viewed as the major part of a research plan. Most of the contents are concerned with numerous details that must be decided on and integrated into a coherent whole. One decision concerns the most beneficial time to formulate a plan. Because it is devised to assist the researcher in organizing all aspects of the project, a plan should be developed as soon as possible. On the other hand, a plan should also clarify and integrate detailed definitions and procedures that depend on field experience. These two somewhat contradictory goals make it difficult to specify an optimal time that is best for preparing a plan. In essence, preparing a plan is an iterative process that modifies preliminary expectations with field verification.

An embryo of a plan probably develops in the mind of a researcher when a problem is first encountered. These preliminary thoughts may be expanded by studying more about the problem, pondering on various design options, and considering a potential study area. After gaining some initial insight into the phenomenon and its spatial setting, the researcher needs to organize ideas about variables to be measured, problems for obtaining data, and probable methods for eventual analysis. The expression of these inputs in written form will constitute a preliminary research plan.

The preliminary plan should serve as a base for seeking suggestions and making modifications. By consulting with other persons who have an interest in the problem itself or with research methodology per se, the researcher can normally accumulate several suggestions. Sometimes the very process of explaining ideas to another person helps in discovering gaps and the need for revisions and additions to the preliminary plan.

Additional modifications may occur during the period of reconnaissance. As discussed previously (Chapter 2), one of the primary objectives of the field reconnaissance is to check on the appropriateness of operational definitions, the identification of field units, the delineation of the study area, and the system for measuring locations. Also trial runs may demonstrate that the initial plans for measuring the variables, generating a sample, or collecting the data may not be feasible. Such a discovery may require major modifications to the research plan. Although the researcher may be discouraged by having to make alterations in a plan that has already consumed much preparatory time, the situation is much better than not prechecking the research design and, thus, not discovering its inadequacies until the gathering of data has commenced.

By the time a researcher completes the reconnaissance stage, the research plan should be in its final form. This is not to imply, however, that subsequent modification will not occur. In fact, most researchers, including scholars with many years of research experience, often encounter a few unanticipated situations or gain additional insights that then require additional procedural definitions. Nevertheless, the researcher should have considered all the major aspects of a research project and be able to provide a realistic plan by the time the goals of the reconnaissance have been accomplished.

Form of Plan

Any organizational format that reminds the researcher of the many facets that must be considered and integrated is satisfactory because the research plan is basically a working guide for the researcher's own use. The two common forms are an outline and a written proposal.

An outline, which requires only a minimum of discussion, can be used for an oral presentation to others such as colleagues in an academic institution, citizens of a community, or members of a funding organization. One of the advantages of an outline is the ease of scanning an entire plan to spot omissions. Another advantage is the speed in making modifications.

A written proposal of research is a description of a plan for conducting research, usually prepared for communication to other persons interested in the project. If the project is undertaken for a collegiate thesis, professorial approval of a research proposal is often required. If the research aims at solving a community problem, the proposal may be presented to a governmental agency, a volunteers' organization, or some other local group that will sponsor the project. If the researcher aims to secure financial support from a funding organization, a formal proposal normally constitutes the major part of the application for funds.

Elements of a Research Plan

Because a research plan describes the way a research problem is to be solved, its elements are essentially the stages of the research cycle (fig. 1.3). These stages are introduced at the beginning of the text and various aspects are mentioned elsewhere so only a summarizing outline is given here (Table 10.1). It is hoped that the reader will recognize in this topical listing many of the basic issues that confront anyone who attempts to solve research problems in a sound and scholarly manner.

Table 10.1 Elements of a Research Plan

A. Statement of the Problem
    1. Core of research problem
    2. Secondary issues
    3. Research phenomena
B. Objectives
    1. General or specific
    2. Review of literature
    3. Justification
C. Relationships
    1. Hypotheses for generalization
    2. Principles for application
    3. Variables
D. Spatial Setting
    1. Study area
    2. Locational system
    3. Identification of field units
E. Data Collection
    1. Techniques
    2. Measurements
    3. Sampling design
F. Data Analysis
G. Budget

Statement of the Problem

In written proposals the statement of the problem is normally a simple sentence that explains the core of the research in a concise manner. It basically answers the query, "What are you doing?" The statement should be concise with the words chosen carefully, even though it is often difficult to express complex ideas briefly. It should result in a statement that communicates the primary message to prospective readers in an efficient manner. The process of selecting the most appropriate words is also beneficial to the writer. This is because it forces the researcher to concentrate on the most essential reason for undertaking the proposed project.

Occasionally a project consists of a complex of problems. Sometimes it is nearly impossible to communicate the nature of proposed research in one sentence. In this case the core statement may be reinforced by a few subsidiary statements that present secondary issues. Clarity in communicating the relative importance of the research components is usually achieved by listing the subsidiary statements in a parallel format directly below the core statement.

The phenomenon being studied obviously will be stated in the core statement of the problem. Even so, as stressed in Chapters 1 and 2, it is necessary to specify an operational definition of the research phenomenon so it can be identified precisely and the results of the research are interpreted correctly. Such an explanation may fit logically into the section on "Measurement" (topic E-2 in the outline, Table 10.1), but some scholars prefer to provide a preliminary clarification near the beginning of the proposal.

Objectives

The section on specific objectives answers the question, "Why are you doing it?" Of major importance, of course, is whether the methodological objective is to provide a solution in terms of *general* principles of location or a *specific* applied case. As discussed above, this is a critical distinction because the objective will affect the overall research design as well as various details.

If the objective is to expand knowledge about the phenomenon by developing generalizations, then the researcher must become thoroughly familiar with what is already known about the phenomenon and its locational characteristics. Such familiarity is normally acquired by reading professional literature, including previous research reports. In a formal research proposal, a summary of the present state-of-the-art should appear in a section that reviews the literature.

If, in contrast, the objective is to apply knowledge toward the solution of a specific problem, the literature review may be characterized by more descriptive materials. Even so, previous studies providing general principles of location that can be deductively applied to the specific case should be included. The review aids the researcher in planning by clarifying the context within which detailed procedural decisions must be made.

The "why-are-you-doing-it" question should be addressed directly in a section sometimes called "justification". For a study intended to expand geographic knowledge about principles of location, the review of the literature should be presented in such a manner that the gaps in, or "frontiers" of, the body of existing knowledge are apparent to the reader. Then the researcher can demonstrate that the proposed study will fill a gap or expand a frontier. Alternatively, for an applied study, the justification normally describes the background that gave rise to the specific problem and the kind of benefits expected to result from its solution.

The justification and objectives are important for convincing others (e.g., academicians, local citizens, or a funding agency) of the merits of the research project, but this section is also helpful to the researcher in planning subsequent elements. This is to prevent gathering a mass of data that only seem to be related to the problem, a practice that unfortunately occurs too often in poorly prepared projects. When a justification of the research is clearly specified, the researcher can best assess the most critical facts that need to be collected.

Relationships

Research that attempts to develop principles of location should list the hypotheses in a manner that exposes the variables and their relationships with the research phenomenon. In many respects the hypotheses will be formulated in such a manner that they can be tested through inferential statistics. In other instances, the "hypotheses" may consist of less precise speculations about relationships that the researcher expects to find. Irrespective of their structure, explicitness, and statistical testability, the hypotheses should contain sufficient detail to specify the kind of data that need to be collected.

If, in contrast, the goal is to apply previous generalizations to a specific problem, the researcher must elucidate the appropriate principles. By including these in the research plan, the researcher has a checklist specifying the data that will be needed. It is clear, therefore, that both hypotheses to be tested and principles to be applied, which may be intermixed in some problems, provide a valuable guide to the inventory of data to be collected.

Some researchers find it useful to enumerate the variables separately. In some proposals, therefore, the related phenomena are listed and appear in the hypotheses and stated principles. The list may contain a general name, an operational definition, and the technique for obtaining data associated with each variable. Such a list is especially helpful in planning research projects when data must be acquired for numerous variables.

Spatial Setting

Selection of a study area that is representative of a larger area is a critical decision when results of the research are to be generalized (see Chapters 2 and 4). The research plan, therefore, should discuss the characteristics of the study area that typify the larger region. The same kind of comments should clarify the time period to which the areal situation applies.

In contrast, if the research applies to a specific problem, the study area is usually already defined. Nevertheless, when writing a proposal, researchers may describe the characteristics of the study area for the purpose of conveying the context of the research.

A decision inherent in any spatial problem concerns the locational system. Although this element has not always been stated explicitly in past proposals, researchers are in actuality forced to make a decision, even if they do not consciously do so. A more advisable policy, therefore, is to include this item in a research plan and to make a rational decision about the measurement of geographic positions.

A similar kind of decision concerns the identification of field units. Choosing appropriate field units is only one of many subjective decisions required in solving a research problem. To achieve a goal of acquiring reliable data, researchers must make those decisions according to objective procedures. This is no less true when identifying the basic areal expression of the individual, either as a discrete object at a place or a minimum areal unit of a continuous phenomenon. The procedure for making these field identifications must be considered prior to data collection and, hence, should be an important element in the research plan.

Data Collection

As stressed throughout this text, several different techniques can normally be used to produce data about the research phenomenon and variables. The advantages and disadvantages of these alternatives should be evaluated by considering the kind of questions listed in "A Guide for Evaluating Field Techniques" (Table 5.2). Most researchers by the time they have completed the final modifications to their research plans are able to specify the field technique most appropriate for collecting each set of data.

Closely associated with the choice of techniques for data collection is the decision on method of measurement (Chapter 3). Because of the many ways various attributes of phenomena can be measured, it is wise to decide on appropriate strategies as soon as possible. Sometimes measurement decisions cannot be finalized until after a pilot study has been completed. Nevertheless, the prudent researcher indicates various options even in early versions of a research plan.

When techniques involve asking questions through a schedule or questionnaire, the actual set of questions and procedures for eliciting information from respondents should be part of the research proposal. In a formal proposal that is to be evaluated by other scholars, the schedule or questionnaire, which may be attached in an appendix, should be available so the instrument of measurement can be judged for its probable validity and reliability. Subjecting the schedule or questionnaire to public scrutiny often reduces the likelihood of creating misleading responses.[2] Even in a less formal proposal, inclusion of the survey questions aids the researcher in a self-assessment of how well the data to be collected may coincide with the type of information that is needed.

Most research projects depend on data collected from a sample, so the manner of selecting the sample is critical (Chapter 4). The sampling design and its representativeness must be considered and incorporated into the research plan. The description of the scheme should contain enough detail for judges of the proposal to evaluate its merits and for the researcher to perform the sampling procedure in a routine manner.

Data Analysis

The nature of the research problem and the wording of the hypotheses strongly influence the way the data should be analyzed. Sometimes an amateur researcher is tempted to collect data and then "see what they show." A far superior strategy is to create some hypothetical data or use the results from a trial run to examine various ways of organizing, testing, and presenting the conclusions from the data. In other words, the researcher should plan how the data will be analyzed before they are collected. This tends to insure that the data-collecting techniques are compatible with the analytical methods.

Budget

Another element that should be considered before plunging into a major data-gathering project is the cost. The most expensive items must be tentatively estimated when alternative techniques are being considered because they can affect the techniques actually chosen. When preparing a budget with more precise numerical values, these same items will reappear along with other expenditures. The budget usually includes costs other than those spent just on collecting data (e.g., computer time for analysis). The amount of budget detail may depend on the purpose for preparing it. A sketchy budget may satisfy the researcher's own appraisal of personal costs, while a funding organization may require precise accounting of each estimated expense item. In the latter case, the researcher is normally guided by a particular budget form.

Evaluation and Applications

Evaluation

Even if a researcher develops a detailed research proposal, this will not guarantee an absence of trouble in collecting data. Many phenomena, especially humans, are too complex for anyone to anticipate all facets of data collection. Researchers who neglect to formulate a plan, or prepare an incomplete one, before attempting to collect field data run a higher risk of frustration and failure.

Application

There are many sources for additional reading on the topic because several books dealing with research methods discuss the task of developing a plan or design.[3] To learn more about preparing a plan, you should try designing one yourself. As suggested for several other aspects of field work, let experience be your teacher.

Notes

1. This message is made very clear by M. R. Hill, "Positivism: A 'Hidden' Philosophy in Geography," in M. Harvey and B. Holly, eds., *Themes in Geographic Thought* (London: Croom Helm, 1981), pp. 38–60.
2. The definition of responses that are "misleading" is related partly to the methodological and philosophical issues of "objectivity" and partly to the motive for generating responses. As stressed in Chapter 6, some polls are conducted and publicized for the purpose of convincing an audience of the supposed popularity of a particular viewpoint. When the motive is to deceive, the questions that produced those results (and/or the sampling design) are rarely included.
3. R. W. Durrenberger, "Identifying a Problem and Developing a Research Plan," in *Geographical Research and Writing* (New York: Thomas Y. Crowell, 1971), pp. 22–35; L. L. Haring and J. F. Lounsbury, "Formulation of a Research Design," in *Introduction to Scientific Research,* 2nd ed. (Dubuque, Iowa: Wm. C. Brown Company Publishers, 1975), pp. 19–26; J. F. Lounsbury and F. T. Aldrich, "Field Research Design," in *Introduction to Geographic Field Methods and Techniques* (Columbus, Ohio: Charles E. Merrill, 1979) pp. 121–137; G. R. Smith, "How to Write a Project Proposal," *Nation's Schools,* 76 (1965), pp. 33–35, 57.

Appendix A

Making a Map with a Plane Table

Constructing a map in the field by using an alidade and plane table begins with a blank sheet of paper attached to the drawing surface of the table. The map is initiated by marking two base points (say A′ and B′) whose separation and relative positions correspond to two ground positions (here labeled A and B respectively) at an appropriate scale and location within the study area. The appropriateness of scale is determined primarily by the comparative size of the mapping sheet and the study area. The positions of the points within the map area should approximate the relative locations of their corresponding ground positions within the study area.

Field mapping commences by planting the plane table at A, placing the alidade along the line A′B′, and turning the table with attached map until the A′B′ line-of-sight coincides *exactly* with its corresponding AB line (fig. A.1). Then, without moving the plane table (even a slight shift in map orientation will cause errors), the alidade is moved so another field position, say C, that can be sighted from A. If the alidade touches point A′, then a penciled line along the edge of the alidade will form a ray from A′ toward a position that will become C′. In a similar manner rays can be drawn toward other visible field objects (e.g., D and others not shown in fig. A.1).

The researcher next carries the plane table to field position B, anchors the stand firmly, places the alidade along line B′A′, and turns the table/map so the B′A′ line matches the BA line of sight. If the B′A′ line coincides exactly with the BA sighting line, then the map is oriented the same as it was at position A. Now, rays drawn from B′ toward C and D will intersect those drawn from A′ (fig. A.2). The intersections determine points C′ and D′ in the manner that a side and two adjacent angles always geometrically determine a triangle, i.e., by triangulation.

All the field positions toward which rays were drawn from A/A′ may not be visible from B. This does not matter providing at least one position (e.g., C) is visible and is mapped from both A and B because this one intersection shows the isomorphism between the map (C′) and earth (C). This means the researcher can move the plane table to position C because its corresponding map location is known, and therefore can repeat the triangulations and extend the map data. This procedure, namely, drawing rays from known field positions to produce map intersections and then moving the plane table to those places that correspond to these new intersections, is one that can be continued as long as necessary to construct a base map.

If all field positions to be mapped are visible from A and B, then theoretically there is no need to move to a third field location because the pairs of intersecting rays from the two points adequately determine all other places. However, it is wise to move to a third field position and draw rays to map locations already identified for the purpose of verifying the accuracy of the previous work. If the third ray passes through the intersection point of two previous rays, this reveals an accurate map. If the rays from three positions do not intersect in a single point, then the triangle of intersections reveals something about the inaccuracies in the map. Getting three rays to intersect in a common point is especially difficult when two, or all three, are almost parallel; thus, the map-maker should generally avoid relying on rays that intersect at a very small angle. Instead, rays that meet at larger acute angles should be used, possibly by temporarily adding a stake in the field that will allow the researcher to establish the locations of the desired positions.

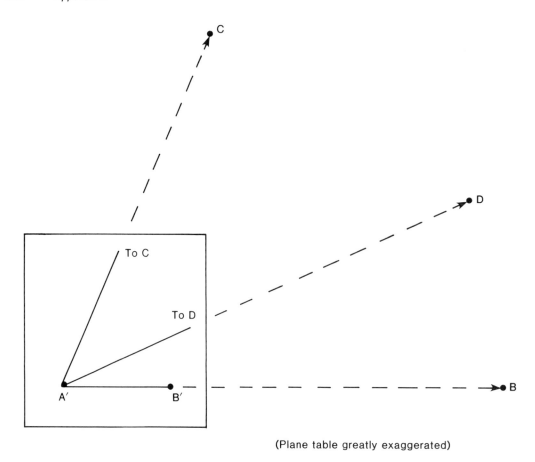

(Plane table greatly exaggerated)

Figure A.1. Mapping from a plane table, position A

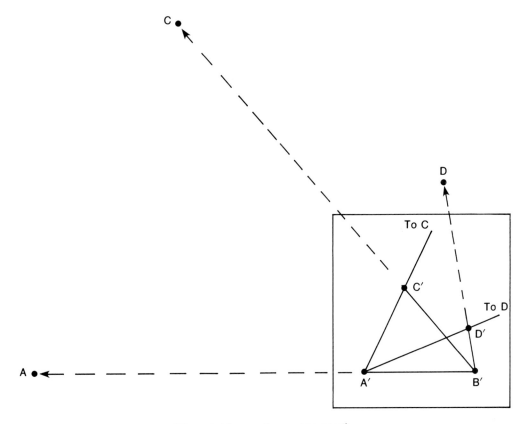

(Plane table greatly exaggerated)

Figure A.2. Mapping from a plane table, position B

Appendix B

Table B.1 Table of Random Numbers

42614	29297	01918	28316	25163	01889	70014	15021	68971	11403	95294	00556
34994	41374	70071	14736	65251	07629	37239	45305	07521	61318	31855	33295
99385	41600	11133	07586	36815	43625	18637	59747	67277	76503	34513	37509
66497	68646	78138	66559	64397	11692	05327	16520	69676	11654	99893	82162
48509	23929	27482	45476	04515	25624	95096	68652	27376	92852	55866	67946
15470	48355	88651	22596	83761	60873	42353	79375	95220	01159	63267	84145
20094	98977	74843	93413	14387	06345	80854	33521	26665	55823	47641	09279
73788	06533	28597	20405	51321	92246	80088	59589	49067	66821	41575	77074
60530	45128	74022	84617	72472	00008	80890	20554	91409	96227	48257	18002
44372	15486	65741	14014	05466	55306	93128	59404	72059	43947	51680	18464
88618	19161	41290	67312	19325	81549	60365	94653	35075	33949	64281	61826
71299	23853	05870	01119	14413	70951	83799	42402	56623	34442	66847	70495
27954	58909	82444	99005	39663	77544	32960	07405	36409	83232	72461	33230
80863	00514	20247	81759	02181	68161	19322	53845	57620	52606	21032	91050
33564	60780	48460	85558	88448	03584	11220	94747	07399	37408	95362	67011
90899	75754	60833	25983	10622	48391	31751	57260	68980	05339	49712	97380
78038	70267	43529	06318	86225	31704	88492	99382	14454	04504	58275	61764
55986	66485	88722	56736	49767	04037	30934	47744	07481	83828	89514	11788
87539	08823	94813	31900	50816	97616	22888	48893	27499	98748	15472	50669
16818	60311	74457	90561	43852	59693	78212	16993	35902	91368	12120	86124
03991	10461	93716	16894	98953	73231	39528	72484	82474	25593	27767	43584
38555	95554	32886	59780	09958	18065	81616	18711	59342	44276	13025	14338
17546	73704	92052	46215	15917	06253	07686	16120	82641	22820	80217	36292
32643	52861	95819	06831	19640	99413	90767	04235	13574	17200	10875	62004
69572	68777	39510	35905	85244	35159	40188	28193	29593	88627	54127	57326
24122	66591	27699	06494	03125	19121	34414	82157	86887	55087	60311	42824
61196	30231	92962	61773	22109	78508	63439	75363	44989	16822	49739	71484
30532	21704	10274	12202	94205	20380	67049	09070	93399	45547	78626	51594
03788	97599	75867	20717	82037	10268	79495	04146	52162	90286	66692	13986
48228	63379	85783	47619	87481	37220	91704	30552	04737	21031	44071	28091
01291	41349	19152	00023	12302	80783	46780	56487	71857	15957	48545	35247
38384	74761	36024	00867	76378	41605	59849	96169	92784	26340	75122	11724
66164	49431	94458	74284	05041	49807	47670	07654	04921	73701	92904	13141
54155	83436	54158	34243	46978	35482	94304	71803	45197	25332	69902	63742
72848	11834	75051	93029	47665	64382	08105	59987	15191	18782	94972	11598

Table B.2   Sources for Imagery from Remote Sensors

Users Service Unit EROS Data Center Sioux Falls, SD 57198	Map and Air Photo Sales U.S. Geological Survey 345 Middlefield Road Menlo Park, CA 94025
National Cartographic Information Center National Center, Stop 507 Reston, VA 22092	Air Photo Sales U.S. Geological Survey Federal Center, Building 25 Denver, CO 80225
Cartographic Archives Division National Archives & Records Service, GSA Washington, DC 20408	Technology Application Center The University of New Mexico Albuquerque, NM 87131
Geography and Map Division Library of Congress Washington, DC 20540	U.S. Forest Service Department of Agriculture Washington, DC 20250
Cartographic Division Soil Conservation Service, USDA Federal Center Building East-West Hyway and Belcrest Road Hyattsville, MD 20781	Satellite Data Services World Weather Building Washington, DC 20233
Aerial Photography Field Office USDA-ASCS Salt Lake City, UT 84125	Coastal Mapping Division National Ocean Survey, NOAA Rockville, MD 20852
Map Information Office U.S. Department of the Interior Geological Survey Washington, DC 20240	General Electric Photo Lab 5030 Herzel Place Beltsville, MD 20705
	Pilot Rock, Inc. Box AS Trinidad, CA 95570

Table B.3 Bibliography of Bibliographical Aids for Geography

Alexander, G. M. *Guide to Atlases: World, Regional, National, Thematic: An International Listing of Atlases Published Since 1950.* Metuchen, N.J.: The Scarecrow Press, 1971, with 1977 Supplement.

Andriot, J. L., ed. *Guide to U.S. Government Statistics,* 4th ed. McLean, Va.: Documents Index, 1973.

Andriot, J. L., ed. *Guide to U.S. Government Publications.* McLean, Va.: Documents Index, 1980, with quarterly supplements.

Besterman, T., comp. *A World Bibliography of Bibliographies,* 4th ed. Lausanne: Societas Bibliographica, 1965–66.

Bibliographical Services Throughout the World. Paris: UNESCO, periodic since 1950, with monthly supplements in *Bibliography, Documentation, Technology.*

Bibliographic Index: A Cumulative Bibliography of Bibliographies. New York: H. W. Wilson, annual since 1937.

Bibliographie Cartographique Internationale. Paris: Armand Colin, annual since 1949.

Bibliographie Geographique Internationale. Paris: Assoc. de Geographes Francais, annual since 1891.

Brewer, J. G. *The Literature of Geography: A Guide to Its Organization and Use,* 2nd ed. Hamden, Conn.: Linnet Books, 1978.

Chandler, G. *How to Find Out: A Guide to Sources of Information for All, Arranged by the Dewey Decimal Classification,* 3rd ed. New York: Pergamon Press, 1967.

Comprehensive Dissertation Index. Ann Arbor, Mich.: University Microfilm International, annual since 1973, with monthly supplements in *Dissertation Abstracts.*

Conover, H. F., comp. *Current National Bibliographies.* Washington: U.S. Library of Congress, 1955.

Cumulative Book Index: A World List of Books in the English Language. New York: H. W. Wilson, annual since 1898.

Current Geographical Publications. Milwaukee: American Geographical Society, monthly since 1938.

Current National Statistical Compendiums: A Bibliography and Index by Basic Subjects. Washington: Congressional Information Service, annual since 1976.

Downs, B. *American Library Resources: A Bibliographical Guide.* Chicago: American Library Association, 1951, with supplements for 1950–61 and 1961–70.

Geo-Abstracts: A— Landforms and the Quaternary; B—Climatology and Hydrology; C—Economic Geography; D—Social Geography and Historical Geography; E—Sedimentology; F—Regional and Community Planning; G—Remote Sensing and Cartography. Norwich, England: University of East Anglia, six per year since 1966.

Hancock, J. C. *Geographer's Vademecum of Sources and Materials,* 2nd ed. London: G. Philip, 1978.

Harris, C. D. *Annotated World List of Selected Current Geographical Series in English, French, and German,* 4th ed. Chicago: University of Chicago, Department of Geography, 1980.

Harris, C. D. & J D. Fellman. *International List of Geography Serials,* 3rd ed. Chicago: University of Chicago, Department of Geography, 1980.

Index to Maps in Books and Periodicals. Boston: G. K. Hall, 1968, with supplements in 1971 and 1976.

International Bibliography of the Social Sciences: Economics; Social and Cultural Anthropology; Sociology; Political Science. Paris: UNESCO, annual since 1955.

Jackson, E. P. *Subject Guide to Major U.S. Government Publications.* Chicago: American Library Association, 1968.

Kovacs, G. *Annotated Bibliography of Bibliographies on Selected Government Publications and Supplementary Guides to the Superintendent of Documents Classification System,* 5th Supplement. Greeley, Colo.: Gabor Kovacs, 1977.

LeGear, C. E., *United States Atlases: A List of National, State, County, City, and Regional Atlases in the Library of Congress.* Washington: U.S. Library of Congress, 1950, with 1953 supplement.

Lock, C. B. M. *Geography and Cartography: A Reference Handbook,* 3rd ed. Hamden, Conn.: Linnet Books, 1976.

Map Collections in the United States and Canada: A Directory, 3rd ed. New York: Special Libraries Association, 1978.

Martinson, T. L. *Introduction to Library Research in Geography: An Instruction Manual and Short Bibliography.* Metuchen, N.J.: Scarecrow Press, 1972.

Minto, C. S. *How to Find Out in Geography: A Guide to Current Books in English.* New York: Pergamon, 1966.

Monthly Catalog of United States Government Publications. Washington: U.S. Government Printing Office, monthly since 1895, with annual *Cumulative Subject Index.*

National Union Catalog: A Cumulative Author List Representing Library of Congress Printed Cards and Titles Reported by Other American Libraries. Ann Arbor, Mich.: J. W. Edwards, 9 monthly issues per year since 1956.

Table B.3—*Continued*

New Geographical Literature and Maps. London: Royal Geographical Society, semi-annually since 1918.

New York Times Index. New York: New York Times Co., since 1851. Also see indexes for other newspapers, e.g., *Christian Science Monitor.*

Palmer. A. M., ed. *Research Centers Directory,* 6th ed. Detroit: Gale Research Co., Book Tower, 1979, with periodic supplements as *New Research Centers.*

Readers' Guide to Periodical Literature. New York: H. W. Wilson, bi-weekly, since 1900.

Research Catalog of the American Geographical Society of New York. Boston: G. K. Hall, 1962, with supplements.

A Select Bibliography: Asia, Africa, Eastern Europe, Latin America. New York: American Field Staff, 1960, with two-year supplements.

Sheehy, E. P., ed. *Guide to Reference Books,* 9th ed. Chicago: American Library Association, 1976, and *9th ed. supplement,* 1980.

Social Science Citation Index. Philadelphia: Institute for Scientific Information, annual since 1962.

Social Science Index. New York: H. W. Wilson, quarterly since 1907.

Statistical Abstract of the United States. Washington: U.S. Dept. of Commerce, annual since 1878.

Statistical Reference Index: A Selective Guide to American Statistical Publications Other than the U.S. Government. Washington: Congressional Information Service, monthly since 1980.

Subject Guide to Books in Print: An Index to the Publisher's Trade List Annual. New York: R. R. Bowker, annual since 1956.

Toomey, A. *A World Bibliography of Bibliographies, 1964–1974: A List of Works Represented by Library of Congress Printed Catalog Cards.* Totowa, N.J.: Rowman and Littlefield, 1977.

UNDOC: Current Index. New York: United Nations, ten per year since 1970.

Vance Bibliographies. Monticello, Ill.: Vance Bibliographies, several each year since 1958.

Vinge, C. L. & A. G. Vinge. *U.S. Government Publications for Research and Teaching in Geography.* Norman, Okla.: National Council for Geographic Education, 1961.

Walford, A. J., ed. *Walford's Guide to Reference Material,* 4th ed. London: The Library Association, 1980.

Winch, K. L., ed. *International Maps and Atlases in Print,* 2nd ed. New York: R. R. Bowker, 1976.

Wright, J. K. & E. T. Platt. *Aids to Geographical Research,* 2nd ed. New York: Columbia University Press, 1947.

Index

Accompanied walk, 199, 203
Accuracy
    data, 15, 32, 51, 64, 66, 72, 85, 88–90, 155,
        204, 205
    locational (mapped), 32–33, 36, 37, 39, 41,
        201, 204, 223
Applied geography (*see* Geographic studies)
Archival (stored) data, 20–21, 33, 34, 183–195,
        199, 208–209
    (*see also* Content analysis *and* Historical
        data)
Awareness, 116, 117, 118, 122–124, 128, 135,
        195, 201, 204, 205, 207
    Hawthorne effect, 118

Bias, 34, 51, 77–82, 90, 91, 96, 98, 103, 120,
        138, 148, 155–157, 159, 188, 194, 201,
        204, 205, 207, 208
Bibliographic sources, 192, 329–330

Card sorting, 64, 148, 154, 163 n
Classification, 15, 47, 52–54, 55–61, 62–64, 65,
        66, 72, 73 n, 85, 94–95, 147, 148, 149,
        175–178
    aggregating approach, 55, 73 n
    divisional approach, 55, 57
    field, 110–115
    supervised, 177, 178
    unsupervised, 177
    (*see also* Identification)
Coding, 113, 126, 141–142
Content analysis, 142, 185–188, 193–195
Control over
    field conditions, 121–122, 123, 124
    laboratory conditions, 122–123, 124
    variables, 10, 110, 121–124, 134, 135, 136,
        137, 194
Cooperation, 115, 128, 139–140, 142, 154,
        156–159, 160, 161, 195, 204, 205, 207
Cost comparisons
    content analysis, 195
    field classification, 115
    field observations, 123, 202
    field reconnaissance, 35, 42
    films, 207

imagery from remote sensors, 174, 179
interviewing, 142, 160, 161
mail questionnaire, 154, 160, 161
photographs, 125, 208
sample, 85, 88–89, 90, 102
schedule, 160, 161
telephone schedule, 152, 160, 161
tracking, 201

Descriptive geography (*see* Geographic studies)
Detection of errors, 114, 124, 128, 194

Ethical issues, 116, 120, 128, 139, 207

Field, defined, 9–10, 23 n
Field diary, 35, 120, 127
Field observations, direct, 17–20, 109–129,
        200–202
Field reconnaissance, 14, 26, 29, 34–35, 41–42,
        57, 84, 101, 120, 126, 135, 190, 192,
        214, 216
Field surveying, 38–41, 223–225
    alidade, 40–41, 223
    compass, 39
    pacing, 38, 39, 111
    plane table, 40–41, 223–225
    transit, 39, 40
Films, 125, 199, 206–207
    stored video tapes, 183, 192
    (*see also* Imagery from remote sensors *and*
        Photographs)
Free story (*see* Interviewing)

Geographic data, 4, 8–11, 30, 36, 37, 125, 126,
        128, 165, 184–185, 194, 199
Geographic problem, characteristics, 3–6, 22, 217
Geographic studies
    applied geography, 6, 7, 11, 13, 14, 16, 25, 28,
        76, 102, 214, 215, 218, 219
    descriptive geography, 6–7, 76, 102
    principles of location, 6, 7, 11–16, 25, 28–29,
        34, 76, 82, 102, 214–215, 218, 219
    theoretical geography, 6, 7
Geometric triangulation, 32, 39–41, 223, 225